THE
ICE BOOK

RED ⚡ LIGHTNING BOOKS

THE
ICE BOOK

Cool Cubes, Clear Spheres, and Other Chill Cocktail Crafts

Camper English

Photography by
Allison Webber

This book is a publication of

Red Lightning Books
1320 East 10th Street
Bloomington, Indiana 47405 USA

redlightningbooks.com

© 2023 by Camper English

Sixth printing 2023

Cataloging information is available from the Library of Congress.

ISBN 978-1-68435-205-0 (hardback)
ISBN 978-1-68435-207-4 (ebook)

CONTENTS

THE
ICE BOOK

Figure 1.1. HELLO. Clear cubes with citrus peels frozen inside.

1

ICE IN THEORY

Clear ice. It's frozen water. Why all the fuss?

You could say the same thing about pizza, bagels, beer, or any other edible or nonedible items—cast-iron pans, chef's knives, or crystal wine glasses—that we obsess over. We want the very best version of our favorite simple things.

Crystal-clear, high-quality ice is somewhere between food and a tool. Yes, clear ice tastes better (but not much better) than milky-white cloudy ice. It also melts more slowly than cloudy ice: this can be a positive or negative aspect depending on how quickly you want to cool your cocktail. Aesthetically, however, clear ice is drastically superior to the cloudy stuff and elevates the experience of drinking a beverage with it.

Technically, the world's best champagne tastes the same whether you sip it out of a Styrofoam cup or from an elegant crystal flute, but the latter vessel might cause you to slow down and appreciate it more. Good ice can do that for your drink as well.

This book is mostly about how to make clear ice via "directional freezing," a simple trick to force water to freeze into ice from one direction to another rather than from the outside in, as in a typical ice cube tray. When we use a directional freezing system, mimicking how a pond or lake freezes naturally, the ice tends to form in perfect clarity as it pushes trapped air and impurities away from the point of freezing.

I didn't create directional freezing, but I did help translate the natural process to the home freezer. More than a decade ago, after many months of experimenting with making clear ice, I figured out that you can imitate how a pond freezes by putting water into a hard-sided beverage cooler, leaving the top off, and placing it in the freezer. Back in 2009, I shared my technique, and although top bartenders around the world began adopting this method to make clear ice cubes, it wasn't until YouTubers and Instagrammers began posting how-to videos of the process that it really took off.

In the years since my blog post, an army of ice nerds has joined me in finding beautiful and fun new ways to make ice in a variety of shapes, bedazzling it on the inside and out, and generally increasing the awesomeness of everyday ice cubes. So I can't take credit for all of the good ideas presented here: just the one Big Idea that inspired the rest. It is my pleasure to share with you some of my favorite manifestations of frozen water.

An Ice Odyssey: How I Figured Out Directional Freezing

Do you remember being told that you should boil the water to make clear ice? When I was a kid, there was even an educational segment between cartoons on TV that told us this. It was science!

But it was all a lie. The boiled water urban legend is something everybody "knows" and yet very few people have fact-checked.

In the early days of the post-2001 craft cocktail renaissance, some bars began making larger ice cubes than the ones made in ice machines to put into slow-sipping drinks like the Old-Fashioned. They froze water in large containers and then cut up the blocks into big cubes by hand. (This was before you could buy 2-inch silicone ice cube trays.) The ice was not clear, but bartenders tried to improve it by boiling the water first. Others melted down small clear cubes from an ice machine with hot water and refroze them.

I heard about this quest for clearer ice and decided to do some home experiments. I boiled water and then froze it. The idea is that if you boil the water all the air inside the water will bubble off. It doesn't work, and the resulting ice is still cloudy. Thinking that perhaps the water was reabsorbing air when it cooled down after boiling, I tried putting airtight lids on containers of boiled water and froze that, but this also produced cloudy ice.

Then I tried letting ice melt and freezing it again to see if the clarity would improve. I repeated this with the same water-filled container every day for nearly two weeks. It didn't get any clearer. Next, I compared frozen distilled water with tap water. Nope. I even froze carbonated water just in case that would magically make clear ice. It does not.

Throughout these experiments I used many different shapes and sizes of containers in which to freeze the water. I started to notice that the ice near the outside of each container was clearer than that near the middle. This was true in round vessels like deli soup containers and in flat square ones like lasagna pans. And it didn't seem to matter whether the water was boiled, melted, or distilled.

I figured out that if you wanted to harvest just the clear parts of the ice, you could use a wide, flat container to make the ice and then cut off the clear outside edges. Further, I noticed that if you didn't let the container of water freeze all the way solid, the outside shell of ice was clear, while the center was still (also clear) liquid. It was only when the middle froze solid after the outsides did that the cloudy part appeared. It seemed to me that the cloudy part of the ice was due to "trapped" air in the middle of the ice block. It is the same in ice cubes made in a tray—the outsides are usually clear while the centers are cloudy.

I tried to release that trapped air in the center liquid so that I would have only clear ice. I inserted a metal straw into the middle of a water-filled container so that the air could escape as the water froze. Instead, the water inside the straw also froze and blocked the air from exiting. I tried a few other configurations, but I couldn't figure out a way to release the air from the middle of a block of ice while it was in the process of freezing.

I theorized that if there were no "middle" of the ice block, then the air couldn't be trapped there. So I tried freezing water in very thin layers, but there was still a cloudy middle to each layer of ice. I ended up with thinly striped ice with white and clear layers. (This can look interesting on its own though—see fig. 1.2.)

I knew I was stuck with cloudy ice. But then I had the thought: If we can't *eliminate* the cloudy part of the ice block, could we at least *relocate* it? The last part of the ice to freeze, the center, is where the cloudy ice forms. But if I used an insulated container, I might be able to control which part of the ice froze first and last.

In theory, I could use an insulated cooler, put it in the freezer filled with water, and leave the top off. As the cooler is insulated on the bottom and sides, the water inside it would freeze only from the exposed top surface downward toward the bottom, then the last part of the block to freeze would be on the bottom of the cooler rather than in the middle. There would still be a cloudy part down at the bottom, but it might be easy to separate from the clear part.

I first tried a soft-sided insulated picnic bag cooler in my freezer, leaving the top open so that it would encourage freezing from the top down. It worked! The top part of the block was perfectly clear and only the bottom was cloudy. But as the bag was made of a flexible material, it expanded as the water froze, and I struggled to remove the ice.

I bought a small, hard-sided cooler to solve that problem and made the same type of clear-on-top ice block. In December 2009 I posted the results on my website. I had discovered the practical way to make clear ice at home. It became known as "directional freezing."

With this technique, people at home could harvest just clear ice and leave the cloudy part behind. Over the next months and years I was able to refine the technique and figure out some of the nuances. But to be honest, I didn't understand exactly *why* water froze into ice this way. It was all based on practical observation.

I then read a terrific book: *Ice: The Nature, the History, and the Uses of an Astonishing Substance* by Mariana Gosnell. In the first chapter, the author watches a lake freeze and consults with scientists about what is happening. I'll skip most of the details because it's surprisingly complicated, but the author describes the way in which water turns into ice and how it behaves while in the process of freezing.

When fast-moving water molecules slow down and form solid ice crystals, the crystals reject air and impurities due to their size and structure, and those impurities and air get pushed away from the point of freezing. Lakes that freeze slowly are clear, as pure ice forms

Pomegranate Margarita on Striped Ice

2 oz (60 ml) 100 percent agave tequila
1 oz (30 ml) triple sec
1 oz (30 ml) lime juice
.75 oz (22 ml) pomegranate juice

Shake all ingredients with ice and strain over striped ice in an Old-Fashioned glass.

Garnish: lime wheel.

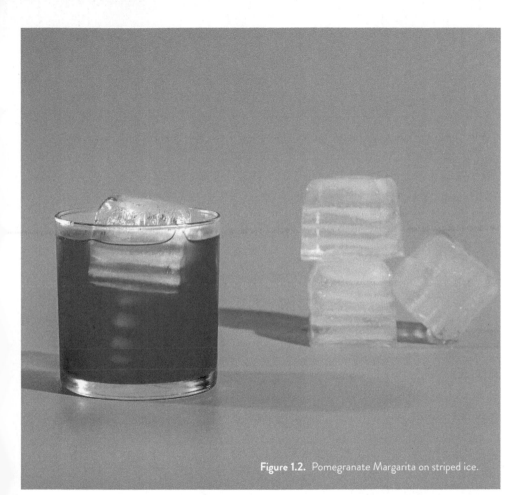

Figure 1.2. Pomegranate Margarita on striped ice.

on the surface of the water and grows downward. The air and impurities in the water are pushed below the forming clear ice.

When water freezes fast, though, the crystals grow quickly and surround some of the air and impurities in the water, thus trapping them inside. Lakes that freeze very fast tend to be covered in cloudy ice with lots of trapped air and impurities. So in order to make clear ice at home, I knew I'd want to control not only the direction of freezing but also the speed.

I understood how to make clear ice and also *why* I was able to do so. And I now knew how to maximize ice clarity. This allowed for a whole range of further experiments, including the writing of this book more than a decade later.

Professionally Produced Clear Ice

Ice sculpture blocks are clear too: do they use giant hard-sided lunch coolers to make them? Well, it's something like that. Ice blocks are frozen in a big box from one side to the other, although in this case the freezing is from bottom to top rather than top down.

Machine-made blocks for ice sculptures are produced in horizontal boxes about the size of a big chest freezer. They are insulated and have a cold plate on the bottom, which allows water to begin freezing from the bottom. Near the top surface of the water, there are water pumps that keep the water in motion as the ice is freezing. This is crucial, as it prevents a layer of ice from forming on the surface of the water. If that happened, then water would freeze from bottom to top as well as from top to bottom, and a cloudy core would form just like in an ice cube tray.

Instead, the water pump prevents the top from freezing over, and the block freezes completely clearly from the bottom up. After a few days of freezing, the last bit of unfrozen water at the top surface of the ice block contains all the minerals and impurities, and the "trapped" air of standard ice has fizzed off. Ice carvers suction off that surface water with a wet/dry vacuum before the block is "harvested."

Probably the most commonly used sculpture ice block maker in the US is made by Clinebell. The machine is nearly 4 feet (1.2 meters) square in total size and produces two 300-pound (136 kg) blocks that measure 20 by 40 by 10 inches thick (50 × 100 × 25 cm) from 40 gallons of water (150 liters) every 3–4 days.

To harvest the block, a hoist like the type that lifts a car engine is used to raise the block up out of the machine and set it onto a surface. Ice sculpture carvers then hack up the huge blocks with chainsaws and chisels.

From Big Blocks to Cocktail Cubes

The big 2-inch (5 cm) ice cubes and tall spears for highball glasses you may have seen at craft cocktail bars are also made from 300-pound ice sculpture blocks rather than in some super-sized ice cube machine. Those huge blocks are cut down into slabs and then again into cubes, spears, and other shapes using a band saw—the type of power saw with a blade that protrudes up from under a table. Then these newly cut cubes are stored in freezers and sold to bars.

Some larger bars and bar groups own ice sculpture block machines and cut down blocks into cubes as part of their prep work. Other bars order whole blocks or smaller slabs of ice delivered to them, and then the bartenders further cut them down into cubes. This all seems like a lot of effort, but the process goes quickly for people who have done it a few times and have the equipment. A lot of bars just buy the finished big cubes from specialty cocktail ice providers. (There were only a few of these cocktail ice companies back in 2009; now most major cities have one or more.)

It would be a lot easier if standard ice machines made these same large cubes so that nobody would have to use a saw, and it turns out that some machines now do. The Hoshizaki company produces machines that make a variety of shapes and sizes of ice including 2-inch cubes and spheres the size of table tennis balls, but these are rarely seen in the United States as of now. In the meantime, the LG company has released a home freezer that produces a few "Craft Ice" ice spheres each day.

In the coming years, we will likely see more machines producing more shapes and

Aperol Spritz over a Clear Ice Spear

3 oz (90 ml) prosecco
2 oz (60 ml) Aperol
1 oz (30 ml) soda water

Add all ingredients to a highball glass with a clear ice spear. (Traditionally served in a large wine glass with ice cubes.)

Garnish: orange slice

Figure 1.3. Aperol Spritz over a clear ice spear.

sizes of big clear ice for the home and for bars, as people have become more familiar with gloriously clear extra-large ice in their drinks in craft cocktail and hotel bars. This book is for those who don't want to wait.

How to Use This Book

This is designed to be a how-to book as well as a source of inspiration for future icy exploration. There are instructions for making clear ice in larger and smaller formats: a block or slab, cubes, and spheres. There are also instructions for adding things inside and outside of those shapes of ice. Finally, we look at colored and flavored ice. I hope that you won't limit yourself to what you see on these pages but be motivated by them to create more fantastic ice in shapes, sizes, colors, and flavors beyond my imagination.

The great thing about an icy arts and crafts hobby is that it costs very little, at least until you cross the threshold of obsession like I have. But still, my little ice projects are far less expensive than say, sport fishing or collecting antiques. Also, you can drink both your successes and your failures—or just water the plants with them.

The cocktails in this book are also mere suggestions of drinks that work with the type of ice demonstrated in each section. They are mostly classic recipes, as there is no shortage of other recipes available online. That said, there are so many terrible drinks websites optimized for SEO and not for quality. My go-to websites for accurate cocktail recipes, particularly when I want to find standard, classic recipes, are Liquor.com, DiffordsGuide.com, and ImbibeMagazine.com.

Today there is an army of ice nerds who compete to show off the most spectacular clear cubes, spheres, and all sorts of ice shapes, as well as ice with all sorts of objects frozen into it. For inspiration, follow the #clearice and #directionalfreezing hashtags on Instagram and other social media apps.

I found a lot of ideas there, too: many of the techniques in this book I developed independently, but many I've gleaned from other ice explorers. I think it is fun to learn from and build on the work of other people in return.

So, for any of you who didn't skip over this section in the first place: Let's get ready to make some ice!

Figure 2.1.
Clear ice slab.

2

HOW TO MAKE
CLEAR ICE

Why Is Ice Cloudy?

In the introductory sections we addressed directional freezing and how clear ice is made in professional sculpture block machines and at home. We're going to cover the home part again in detail because the rest of the book—and your success at making great ice—depends on it.

Consider a typical ice cube, made in a typical ice cube tray in a typical freezer. The cube is cloudy but not completely. The cloudy white parts are mostly near the center of the cube, while most of the outsides are clear. Why is that? And what's in that cloudy part anyway?

Water doesn't turn solid all at once when it freezes. A lake freezes from the top down because the cold air is at the surface, but in most ice cube trays, cold air surrounds each cube on all sides of the tray. It freezes from the top, the bottom, and the sides in toward the center. When the water freezes, its molecules slow down and lock into place, forming a solid crystal lattice. As that happens, anything that is not pure water gets pushed away from the point of freezing.

Because the ice cube in the tray is freezing from the outside in, all the nonwater impurities get pushed in toward the center of the cube—the last part to freeze. And that's where the cube is cloudy.

So, what are those impurities? They're the minerals like sodium, calcium, and magnesium in your tap water, possibly chlorine, and any other organic matter. They get pushed toward the center of the cube. But all these impurities are less important than the biggest source of cloudiness: air. Air is naturally dissolved in water, and as the cube freezes, the air is pushed together toward the last part to freeze, where it forms pockets. These little pockets have big ice crystals in them that refract the light passing through the cube, making it look extra white and cloudy instead of clear.

Directional Freezing

In order to make clear ice, we just need to get the trapped air and other impurities out of the water, right? Likely your first thought is to try filtering the water, or using distilled water, to get rid of the minerals and other solids. Then to get rid of the air in the water we need to degas it by boiling the water or sucking the air out with a vacuum. That would seem correct. Unfortunately, neither removing minerals nor using any conventional method of degassing the water will make the ice much less cloudy.

The way we make clear ice in this book is not by getting rid of impurities or cloudiness. Instead of eliminating cloudiness of ice altogether, we control where the cloudy part of the ice—the last part of the water to freeze—ends up. Rather than freezing an ice cube or ice block from the outside in, we make it freeze from the top down toward the bottom.

"Directional freezing" is a method for making clear ice by controlling the direction in which water freezes.

Consider an ice cube in an ice cube tray again. If we insulate the tray on the bottom and sides but leave the top open, the insulation forces the water to *only* freeze from the top down rather than from the outside in. Now as the water turns to ice, it pushes trapped air and impurities to the bottom of the ice cube tray rather than into the middle. The last part of the water to freeze is where the ice will be cloudy, which is now at the bottom of the cube.

How to Take Advantage of Directional Freezing

The practical way to take advantage of directional freezing isn't to insulate a small ice cube tray to make ice cubes that are clear except at the bottom. A better and easier way to make

Figure 2.2. Clear versus cloudy cube. A typical ice cube is cloudy in its core.

clear ice is to use a hard-sided insulated container. For more than ten years now, I've made my beautiful crystal-clear ice using a hard-sided cooler designed to hold a big lunch or a six-pack of soda.

The insulated cooler forces the water to freeze like a pond does. Fill the cooler with water and place it in a freezer with the top off so that the water is exposed to air on the surface. Because the cooler has insulation built into its sides and bottom, the water freezes from the top down rather than from the outside in.

The first part of the water to freeze into ice—the top—is perfectly clear because when it forms its crystal lattice, the trapped air and impurities are pushed away from the point of freezing into the water below. The last part to freeze, the bottom, is where all the cloudy white ice ends up. Then you can chop off the bottom cloudy part of the ice block you've just made and keep only the clear top portion.

Better yet, you can completely avoid making cloudy ice. The trapped air and impurities keep getting pushed away from the point of freezing into the remaining water underneath. As the ice freezes down to near the bottom of the cooler, there is too much air in the water with no place to go. This is when the cloudy ice forms—but only if you let the water freeze that long.

If you remove the cooler from your freezer before that last bit of water freezes, you will be left with *only* the clear top slab of ice. My insulated cooler in my freezer at home takes about four days to freeze solid, so I try not to let it stay in there that long. After two days or so, I pull the cooler out of the freezer and turn it over; the clear slab slides out, and then the unfrozen water at the bottom sloshes out all over my kitchen. But the slab of ice is crystal clear. By allowing just some of the water to freeze into ice, the air and impurities remain in the unfrozen water that can then be discarded.

A Quick Experiment to Confirm These Theories

If you're having trouble visualizing how ice usually freezes, or you just want to run an experiment to prove to yourself that this is sensible, do the following: Fill a small, roundish plastic container like a take-out deli container with water and allow it to freeze for three or four hours. Pull out the container and look at the ice—it probably won't be fully frozen yet, but the outer ice shell will look pretty clear. The inside will probably be clear water that you can hear sloshing around. All clear, so far.

Now put the container back in the freezer to let it finish freezing. The ice will come out cloudy mostly in the middle and those outsides remain clear. As you will see, the first part of the water to freeze is usually clear, and when it freezes the air and impurities are pushed into the center of the block where they are trapped.

Make a Mostly Clear Ice Block, Just to See How It Works

If you freeze an entire cooler full of water, you'll end up with a block that is about 75 percent clear on top with the rest being cloudy. There is no need to freeze an entire block if you only want to make clear ice, but maybe you'd like to do it anyway to see what happens. (What happens is pictured in figure 2.3.)

Keep in mind, though, that water expands when it freezes, so allowing the entire cooler container to freeze can weaken the plastic. If you freeze the entire cooler solid too many times, it will eventually crack. I don't recommend letting the block freeze solid, but likely you'll sometimes forget and let it go too long.

1. Buy or make an insulated container. I use a hard-sided Igloo brand cooler. You can make your own if you are ambitious, although in my experience the professional ones work better than homemade versions. (You can often find used coolers missing their tops in thrift stores.)
2. Make space in your freezer. This might be a challenge!
3. Remove the lid of the cooler, and fill it with regular tap water. You don't need to fill it all the way up. Place the container in the freezer.
4. After a few days (about four days in my home freezer), the ice should look clear on the surface but cloudy at the bottom. (I find it helpful to shine a flashlight on the block to see how much has frozen.) Remove the cooler from the freezer. Tip it over and wait for the ice block to slide out. It might take more than half an hour to do so.
5. If you are impatient (and who wouldn't be?) you can run warm water over the bottom of the cooler in the sink while holding it upside down. You'll probably hear the container make groaning sounds and start to dislodge. It will eventually drop out.
6. If you have chlorinated water, when you separate the block from the container you may smell a big burst of chlorine. That's normal and a sign that the freezing water concentrated not only the air and minerals at the bottom of the block but also the chlorine!
7. Marvel at your ice block. You have mastered directional freezing. See chapter 3 for how to cut off the cloudy part of your ice block and how to cut the remaining slab into cubes.

Figure 2.3. Making clear ice. Freeze water in an insulated cooler with the top off. At the bottom of the cooler, the last part to freeze, the air and other impurities are trapped.

Make a Perfectly Clear Slab of Ice—the Everyday Way

Here we make a completely clear slab of ice that can be cut up into cubes. The process is the same as making a full block of ice as we did in the last section, except we only let it freeze long enough to make the clear section on top as in figure 2.5.

1. Get an insulated container like a hard-sided cooler.
2. Make space in your freezer.
3. Fill the container with regular tap water. You don't need to fill it all the way up.
4. Place the container in the freezer with the top off.
5. After about two days, remove the cooler from the freezer. Tip it over and wait for the ice block to slide out. As it does, the unfrozen water in the bottom will splash out, so be sure you're doing this in the sink or somewhere that can get wet.
6. Cut your slab into cubes and other shapes. (See chap. 3.)

In reality, the cooler isn't perfectly insulated, and ice crystals form not just on the surface of the cooler but grow down its sides and sometimes across the bottom, too. A thin ice shell forms around the inside walls and bottom of the cooler beneath the clear slab, but it is easy to knock that ice shell off with something like a muddler, a wooden spoon, or an ice pick after you remove it from the cooler. Marvel at your crystal-clear ice slab!

Note that it can take some time for the slab to slide out of the cooler. I am often running to work in the morning and don't have the extra twenty minutes to wait, but I don't want to let the block freeze an additional eight hours. If you find yourself in a rush, you can run hot water over the back of the cooler to get the slab to slide out faster; or instead, just transfer the cooler to the refrigerator for the day. Ice slabs and other shapes will last surprisingly long in the refrigerator without melting.

Keep in mind that you do not need to fill the cooler all the way up if you don't want that much ice. If your goal is to make a slab only about 2 inches (5 cm) thick (which is a great thickness for cutting the ice into big cubes), then you only need to fill the cooler about 5 inches (13 cm) deep with water. You'll probably only need two days for it to freeze. This is how I make my ice slabs and cubes at home.

However, if you want to make a larger batch of ice, you might want to fill your cooler nearly full with water and let it freeze for three full days or so. The tradeoff is that thicker ice is more challenging to cut up later.

If you live in a part of the world that experiences freezing outdoor temperatures, you can probably do this (and other ice projects) by setting your coolers and other insulated containers outdoors and letting Mother Nature do the work.

Figure 2.4. Directional freezing. A block frozen in the cooler with the top off.

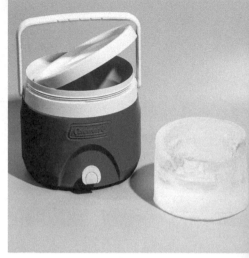

Top left, **Figure 2.5.** Clear slab. Allow a water-filled cooler to freeze for a shorter length of time and harvest only a clear slab of ice.

Top right, **Figure 2.6.** Clear ice punchbowl. Fill an insulated beverage cooler with water and place it in a freezer for a few days. This bowl was frozen in a 2-gallon (7.5 liter) beverage dispenser for three days, then a cavity was hollowed out to make the bowl.

Left, **Figure 2.7.** Negroni Spagliato in a clear ice punchbowl.

Negroni Spagliato in a Clear Ice Punchbowl

1 oz (30 ml) Campari
1 oz (30 ml) sweet vermouth
1–2 oz (30–60 ml) sparkling wine

Scale recipe for the size of your punchbowl. Add all ingredients to punchbowl.

Garnish: orange wheels.

Many people own a second home freezer like a chest freezer, with room for some larger projects. A few bar owners I know have walk-in freezers and have used multiple wide but shallow "Party Stacker" coolers to make large slabs of ice to cut down in bulk.

Another ice nerd I know made a punch-bowl out of a large 5-gallon beverage cooler that he sat inside a chest freezer. Freezing water in the beverage cooler resulted in a big cylinder of ice. He then carved out a bowl shape into the top of the cylinder using an electric chainsaw. I replicated this but with a 2-gallon beverage cooler in my regular freezer (fig. 2.6). I neither have a chest freezer nor a chainsaw, so I melted the bowl shape into the top of the cylinder by repeatedly pouring hot water onto the top to hollow it out.

Make the Best Ice

For the best and clearest ice, your container of water should sit undisturbed at barely below the freezing temperature. That's right, the best ice comes from the warmest freezers! A temperature just below the freezing temperature of 32°F (0°C) allows the water to slowly come to an equilibrium temperature and then change into ice. In theory, if your freezer stayed at exactly this temperature, you might not even need an insulated container to make exceptionally clear ice.

The reality is that freezers cycle on and off when the internal temperature hits minimum and maximum temperatures. My freezer at home varies by as much as 20 degrees Fahrenheit on one setting. The United States Food and Drug Administration (FDA) recommends a maximum temperature of 0°F (−18°C) for food safety, which is far colder than the freezing temperature of water. In my freezer, the warmest I can set it is about the maximum food-safe temperature, so my food won't spoil while my ice freezes as slowly and clearly as possible.

Freezers often have a fan to evenly distribute the cold air, but these can interfere with making great ice. When the fan blows directly on freezing water, it freezes faster—and not usually in a good way. The fan can cause a cloudy bump on an ice slab in a cooler, make ice spheres freeze too fast for clarity, or amplify the "mystery pillar" discussed on page 41.

To avoid the fan breeze, you might be able to move your insulated container in the freezer or make a wind guard to redirect the air from the fan so it's not blowing directly on top of the ice. (Don't block the fan itself but make a wind shield for the ice.) I lean a thin plastic cutting board across the top of the cooler—I find this works to block the wind. Note that if you seal the entire surface of your cooler with plastic wrap, the rate of freezing slows significantly. Ideally you want to allow air flow to reach the top of your cooler, but avoid direct wind from the fan.

Another cause of cloudy ice is when the container gets shaken in some way. Try to avoid jostling the container holding your freezing water: otherwise, the trapped air in the water being pushed down into the bottom of the cooler can form bubbles that float up and stick to the underside of the freezing ice. You may get a small bubble trail or a little puddle-shaped pocket in your ice.

The jostling of the cooler is often caused by people opening and slamming the freezer or refrigerator door shut. (In my house it is caused by me pulling the cooler out of the freezer to check it too often, but this has a lesser impact.)

To repeat, the main causes of cloudy ice in a directional freezing system are excessively cold freezer temperatures, the fan blowing directly on top of the freezing ice, and shaking of the freezing cooler. This information is reiterated in greater detail in Appendix 1 so you can refer to it for other ice projects.

For many ice nerds, "very clear" is not clear enough; they want every single bubble or cloudy patch eliminated. In the next section, we'll look at some refinements.

Small Additional Improvements

In a directional freezing setup, we force trapped air and impurities to move to the last part of the water to freeze. But if we remove the air and impurities from the water in the first place, the ice will freeze with perfect clarity without the need for directional freezing, right?

It makes sense in theory. However, I don't know anyone—including scientists and engineers—who has been able to make clear ice by eliminating air and impurities.

But could removing *some* air and impurities from water before freezing at least *improve* the clarity of ice, or increase the ratio of clear to cloudy ice in a block via directional freezing? I have tried the below improvements, and while they do increase clarity a little, there is still a lot of cloudiness remaining. In my opinion, the added effort, energy, and expense of trying to remove the air and impurities from the water isn't worth the minimal improvement to ice quality in a directional freezing setup. Directional freezing does most of the work.

But I encourage you to experiment!

Getting Rid of Air

Does your kitchen sink have an aerator attachment on the faucet? That little screen is there to infuse your water with air. Try unscrewing it to see if it improves the clarity of your ice blocks.

You can also heat or boil the water to see if that improves your ice. We know that water contains air and that boiling the water causes bubbles to float up and pop. Hot water holds less air than cold water, so warming it up makes sense as a method to reduce air that will be trapped during freezing.

In my experiments, I decided to boil water for a long time to get as much of the air out as possible. I tried boiling the water for up to an hour, and sadly it only produced a steamy kitchen and fairly standard ice. I later learned that the air in water is released during the very first part of boiling. The rest of the bubbles that form in boiling water are from water vapor, not air. So if you're going to try boiling your water, you won't need to do it for very long, just until it reaches a "rolling boil."

I later did an experiment to see if using boiled water was an improvement over using simple tap water. I froze the same amount of boiled and unboiled in a cooler and compared how large the cloudy section of the resulting fully frozen blocks was. I found that boiled water doesn't make much (if any *more*) clear ice in the block—the clear part of the final blocks had the same thickness. That said, the level of cloudiness within the cloudy part of the ice was reduced. In other words, the part of the ice that I throw out at the end looked better.

Using boiled water in the directional-freezing insulated cooler system might slightly improve the clarity of the ice, but it takes a lot of energy to boil the water and a lot of extra time to let that water cool down. Because directional freezing produces plenty of clear ice, I do not bother with that effort. But your water source may be different than mine, or you may be more of a perfectionist than I am. Give it a try and see what happens.

Getting Rid of Minerals and Impurities

The process of directional freezing pushes out air and impurities into the last part of the water to freeze. Thus, the clear part of the ice contains fewer minerals and other impurities—we could say that making clear ice is a type of water purification. If you wanted to improve the taste of your drinking water without using a water filter, you could freeze it in a cooler and then only drink the water from the clear part of the ice.

I haven't found that running the water through a household filter pitcher makes any difference in the resulting clarity of the ice. It seems that most of the cloudiness in ice comes from the trapped air rather than minerals and other impurities. So I don't bother running the water through a filter before freezing it in a directional freezing system, even though I drink filtered water at home and use filtered water in standard ice cube trays.

But if your water tastes really bad, by all means use filtered water for your ice. It certainly won't make things worse.

A Challenging and Impractical Way to Make Great Ice

The directional freezing setup in the cooler forces the cloudy part of ice to form at the bottom of the insulated container. But if we were able to freeze the water in a cooler from the bottom toward the top instead, wouldn't the trapped air just float off the top? Yes, it would, in theory.

You could cut off the outer insulation around the bottom portion of a cooler, fill the cooler with water, and put the insulated top on the cooler. Now the cooler would be insulated everywhere except the bottom and thus freeze from the bottom toward the top. After a couple days you would have a big clear block.

But there's a problem: ice is less dense than water, and so it floats. Plus, just like in our top-down system, ice makes a thin shell around the cooler's interior. Even in this bottom up configuration a layer of ice forms on the top surface of water in the cooler, meaning that air will be trapped and unable to float off the top. The water in the cooler is going to start freezing from bottom to top but also from top to bottom. That means the middle part of the block will be where all the trapped impurities will end up, leaving you with a cloudy center section just like in a regular ice cube.

However, if we were able to do something to keep the top surface water from freezing over in this same system, then we'd have ice only freezing from the bottom up. How about a water pump that keeps the water circulating at the top of the cooler so that it doesn't freeze?

That does work, and that is exactly how sculpture ice block machines like the ones discussed in the first part of the book operate. These machines are essentially big, insulated coolers with a cold plate on the bottom. At the surface of the water, there is a water pump that keeps the water moving and prevents it from freezing over. In these machines, the ice freezes from the bottom up, and when the block is almost fully frozen, the remaining water (that is filled with concentrated minerals and tastes salty) is drained off the top of the block.

Several of my more ambitious readers have built homemade versions of this system. They take a mid- to large-sized insulated cooler, cut the insulation around the bottom so that only a thin inner plastic remains, and insert a tiny aquarium pump near the surface. (Some people do this with a larger camping cooler placed inside a chest freezer.) This works, and it makes a big clear block. In fact, the made in this system with a pump is actually glossy and almost silver in color: it's better than the ice made with top-down directional freezing!

But making ice this way requires a lot of work. The reason I have not replicated this system at home for my ice is because I'd have to cut up my cooler, plus I'd need to put an aquarium pump in my freezer and have the electric cord running out the door. Even for an ice nerd like me, that's a bit more effort than I'm willing to make. That shouldn't stop you from giving it a try though if you're so inclined.

Clear Ice Trays: How They Work

Over the last decade, many readymade clear ice trays have come onto the market. All of them take advantage of directional freezing, using an insulated container to force the water to freeze from the top down just like the insulated cooler we used in the previous section. Some of them are very sleek and expensive, while others are quite basic.

Each product has an ice cube tray that sits inside the insulated container at the top, leaving some extra space below the tray. There are holes in the bottom of the tray so that

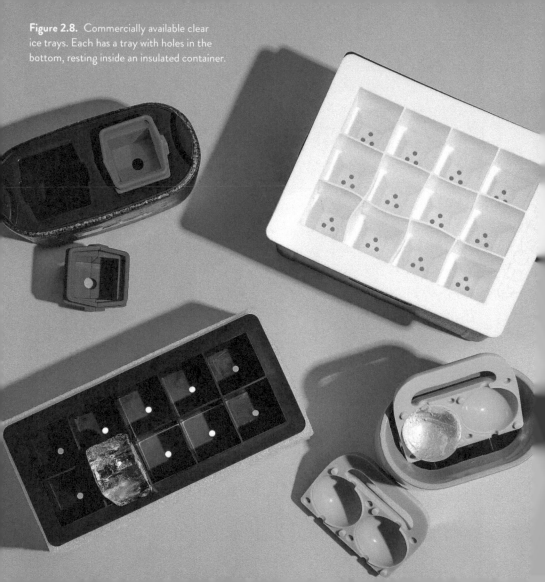

Figure 2.8. Commercially available clear ice trays. Each has a tray with holes in the bottom, resting inside an insulated container.

when the water inside the tray freezes from top to bottom, it pushes the trapped air and impurities out of the holes and into the rest of the insulated cooler water below the tray. The ice that freezes inside the tray is clear, while the cloudy ice forms below it.

Many (but not all) clear ice trays have a reservoir container that sits inside the insulated container, and the ice cube tray sits inside that. This allows for easier removal of the final ice. To extract the clear ice, you pull the perforated tray containing the clear ice off the top of the reservoir.

There is a wide range of clear ice cube products that make anywhere between one and twelve cubes at a time. They do take up a lot of space in the freezer. But on the plus side you don't have to cut up your ice by hand after freezing a slab.

We'll investigate making your own clear ice trays in the next section.

DIY Clear Ice Cube Tray in Three Ways

You can purchase an all-in-one clear ice tray to make clear spheres or cubes, and most of the ones I have tried make great ice. Or you can make your own clear cube tray using an insulated cooler that you already own, plus a silicone ice cube tray.

Method One: Hang a Perforated Ice Cube Tray from the Top of a Cooler

First you'll poke holes in the bottom of a silicone ice tray, and then you'll hang it in your insulated cooler at the top. You can use any kind of hole punch to poke through your silicone ice trays, but ideally you want to make a clean hole rather than just a perforation with a knife. Some people have used narrow metal straws and other hollow metal tubes to do this. I use a "drip irrigation tubing hole punch" (the light blue object pictured in fig. 2.9) after seeing my friend Dave have success with that tool. They cost about $7 online, or you should be able to find them in a gardening store.

Surprisingly, the holes at the bottom of your trays don't need to be that large: I poke a single small hole in the bottom of each compartment of my 2-inch cube trays, and that's enough to make clear cubes. Feel free to poke more holes in your tray, but keep in mind that too many might weaken the integrity of the tray over time.

Now, suspend the perforated ice cube tray near the top of a hard-sided cooler. In figure 2.9, I used a cut-off ruler to help support one side of the tray, while resting another lip of the tray on top of the cooler. Fill the cooler to the top so that the water fills the tray as well as the entire cooler.

Place the cooler with tray inside in your freezer and wait. There is a lot of water in the cooler to cool down, so it might take thirty hours or so for the water to freeze from the top down to just below the bottom of the tray. The only part of the cooler that needs to be frozen solid is the top section where your ice tray is.

When you see that the ice has frozen at least to that depth (admittedly, it's hard to tell how deep the clear ice goes), remove the cooler from the freezer. Dump out the slab of ice containing the perforated ice cube tray. It is usually quite easy to separate the tray from the rest of the block; just hit the ice slab outside the tray with the bottom of an ice pick or another blunt object, and it usually breaks away easily.

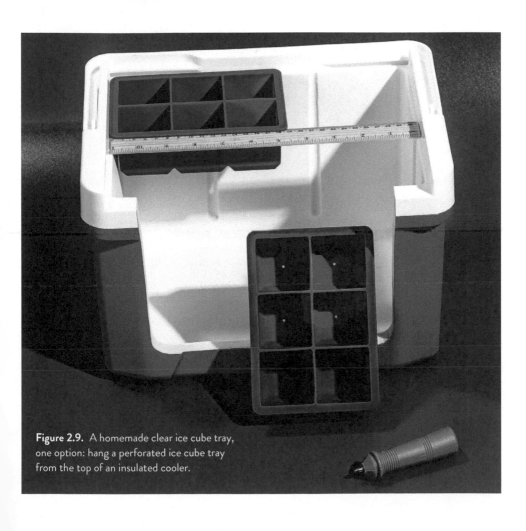

Figure 2.9. A homemade clear ice cube tray, one option: hang a perforated ice cube tray from the top of an insulated cooler.

Though only one cube tray is pictured, you should be able to fit two silicone trays in the cooler at the same time. If you only use one tray, then you will have the remaining part of the clear ice slab from which to cut some highball spears or other cube shapes.

Method Two: Place a Perforated Ice Cube Tray on a Riser in a Cooler

The previous method of hanging an ice cube tray from the very top of the cooler means that the whole cooler is full of water. All that water must cool down in the freezer so that it can begin freezing at the top. That uses up not only a lot of energy but also excess water that will probably be dumped out after we remove the tray. You can save that water, energy, and freezing time by using a much smaller insulated cooler (if you can find one), or by resting the perforated ice cube tray on a riser closer to the bottom of the cooler. That way you only have to fill the cooler with water to the top of the tray.

The riser is necessary to elevate the perforated ice cube tray above the bottom of the cooler—above where the cloudy ice forms. By lifting the tray off the bottom of the cooler, you allow the trapped air to be pushed out the hole at the bottom of the tray.

For the riser, you can use anything that doesn't block the holes in the bottom of the tray. In figure 2.10, you can see I've made the riser from a plastic grid (that was part of a test-tube holder I had lying around) on top of a few pieces of PVC pipe for legs. In other experiments I have used soda bottle caps, a yogurt cup, various plastic boxes, empty cans, etc. The riser doesn't need to be particularly tall nor very sturdy. Ideally though, you want something that doesn't float.

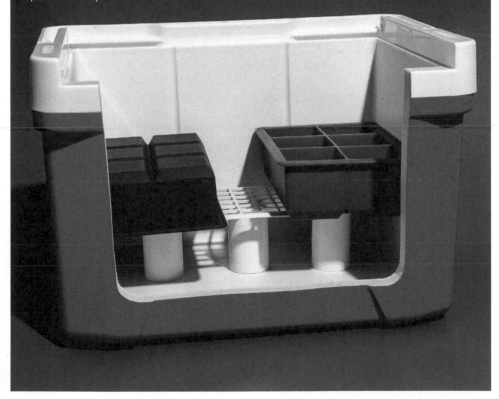

Figure 2.10. A homemade clear ice cube tray, more options: rest a perforated ice cube tray atop a riser, or use a nonperforated tray upside down atop a riser.

This setup works the same as hanging the tray at the top of the cooler. Fill the cooler with water up to the top of the perforated ice cube tray, let it freeze for about a day, then break the tray out of the clear slab. You can also fit two trays next to each other in this system.

Method Three: Place a Nonperforated Ice Cube Tray Upside Down on a Riser in a Cooler

I figured this one out when one of my trays tipped over. Instead of poking a hole in your silicone ice tray, you can turn it upside down, set it on a riser, and then fill the tray with water to the top of the tray (actually the bottom) as in the previous two methods. This is pictured on the left side of figure 2.10.

In this setup, the water is still freezing from the top down, and the shape and separations within the ice cube tray will make individual cubes. If you only let the water freeze for twenty- four hours or so, the water will likely only freeze down to near the bottom (actually the open top) of the tray inside the cooler. Dump the slab out of the cooler, and the cubes still separate easily.

Of course, if you let this setup freeze for longer, the resulting cubes will be stuck to the top of the slab of ice that has formed beneath them. I have done this and found it still made pretty decent cubes that could be chipped off the slab. If you're hesitant to poke holes in your ice cube trays, give this a try—or try the method given in the next section.

The tricky part of this setup is getting all (or most) the air out of the upside-down tray. If you just set the tray on top of the riser, it is likely to float. Instead, submerge it entirely right side up, turn it over underwater, and slip the riser beneath it.

Pretty Good Cubes: Set a Tray Inside Your Cooler

You won't need to poke holes in your silicone ice cube trays or make a riser for this setup; just place your silicone ice cube tray on the bottom of your cooler. Now fill both the tray and the surrounding cooler with water up to the top of the tray. When it freezes, the trapped air and impurities will be pushed down to the bottom of the cooler and to the bottom of the individual ice cube compartments. When you empty the tray after freezing, the bottom of your ice cubes will be cloudy: but *only* the bottoms. This ice looks far better than if you let this same tray freeze outside the cooler without directional freezing.

You can now remove the cloudy bottom of these cubes should you choose to. I think the easiest way to do this is with the bottom of a pan—see figure 3.7. Or leave them be and enjoy better-than-usual big ice cubes in your drinks.

On the "Mystery Pillar"

If you use a clear ice tray that holds six or more cubes, you will probably encounter a "mystery pillar" situation. Some people call it the "sacrificial cube." One of the cubes freezing in the tray pushes upward out of the tray toward the ceiling. It can grow more than 6 inches (15 cm) tall, depending on the tray and cooler you use. It is a very strange phenomenon!

I don't fully understand what's going on, but I believe the mystery pillar has to do with the pressure inside the cooler as water expands into ice. The perforated hole in the bottom of the tray where the mystery pillar forms is like a release valve. Usually, the mystery pillar is very cloudy, which is why some people call it a "sacrificial" cube – you're probably going to throw it out.

Note that if your cooler and tray is near the top of the freezer, the pillar can grow tall enough to hit the interior ceiling of the freezer and potentially damage it. One time I had a tray get jammed into the freezer by the mystery pillar, and it took a herculean effort to remove it. Just check in on your freezing trays at least once a day and make sure to remove the pillar from the tray or the tray from the freezer before the mystery pillar gets out of hand.

Novelty Ice Trays Using Directional Freezing

There are hundreds of novelty ice cube trays, usually made of silicone, available for sale in shapes like letters, numbers, hearts, stars, highball spears, fish, hexagons, flowers, and more. The trays can be used for chocolates, jellies, and even soapmaking and can be found in housewares and baking supply stores. The chocolate or other solid items made in these trays may look great, but the ice that comes out of them is usually cloudy. When water is frozen in these trays, it is often hard to tell what the shapes were meant to be, especially the ones with more complicated designs.

A way to vastly improve the ice produced in these trays is to set them inside a cooler, whether you punch holes in the bottom of the trays or not. If you place these trays in a cooler and fill both the tray and the surrounding cooler with water up to the top of the tray, the ice will freeze from the top down and the cloudy part of the ice will be on the bottom of the shape rather than in the middle. This makes the shapes much easier to decipher. This is the same technique as found in the "Pretty Good Cubes" section above.

Figure 2.11. Novelty silicone ice trays. The ice frozen in them can be made clearer using a directional freezing system, or for perfectly clear shapes, poke holes in each.

Figure 2.12. Godfather cocktail in ice shot glasses. Shot glasses were made in a nonperforated tray in a cooler, causing the cloudy parts of the ice to form only on the top of the glasses.

Godfather in Ice Shot Glasses

2 oz (60 ml) Scotch whisky
1 oz (30 ml) Disaronno

Stir ingredients with ice then strain into ice shot glass.

Garnish: orange twist.

The shot glasses from figure 2.12 were made this way. Typically, these shot-glass shapes would be cloudy throughout the whole shape, but here they are cloudy only on the very tops. (The trays freeze the shot glasses upside down.) The shot glasses are not perfectly clear, but they're a lot clearer than they would be without taking advantage of directional freezing.

You can also poke holes in the bottom of novelty trays, as we did with big 2-inch (5 cm) cube trays to make clear versions of these shapes. I did this with the heart-shaped ice in figure 2.11 and the spears in figure 4.16. This is more successful in trays with flat bottoms than in three-dimensional novelty ice trays like the skull trays elsewhere in this book.

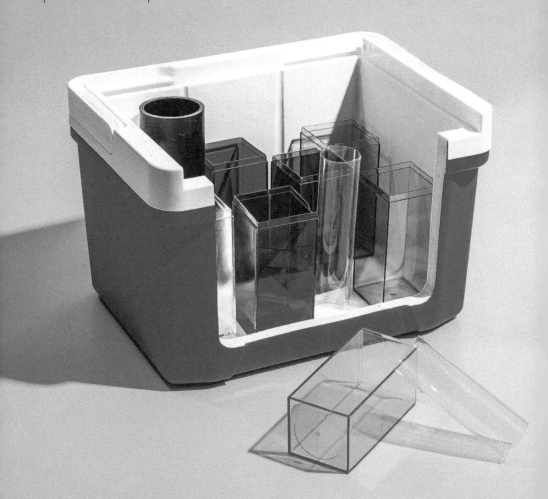

Figure 2.13. Spears made in a cooler with plastic boxes. The set-up.

Figure 2.14. Spears made in a cooler with plastic boxes. The results.

Plastic Boxes to Make Spears in a Cooler

You don't need to limit yourself to silicone ice trays. Long before I learned about the easier method of punching holes in the bottoms of silicone ice cube trays, I found hard plastic (food-safe) boxes for sale and made many different spears and cubes with them. You may not be able to find the same plastic boxes I used here: so consider this a proof of concept for adding other shaped containers to your directional freezing system.

Also pictured in figure 2.13 are a couple of plastic tubes: a small section of a thin poster tube, and a larger wider pipe. These made surprisingly great ice. You might try using plastic takeout containers for food or soups, plastic bottles with the bottoms and tops cut off, or other shapes. You can also use metal tubes and boxes instead of plastic.

Fill the boxes with water as well as the surrounding space in the insulated cooler. As with novelty ice cube trays in the previous section, you can choose to poke or drill holes in the bottom of these trays. Or you can leave them intact. If you do poke or drill holes in the bottom of your containers, set them on a riser so that the clear ice will rest inside the containers, leaving the cloudy ice below them.

Clear Spheres

Everybody loves a clear sphere. You don't need a special excuse to put one in your drink, but a sphere does have the lowest surface-area-to-volume ratio of any shape. This means that spheres will melt slower than any shape of ice with the same volume. Also it looks cool.

You can buy a ready-made clear ice sphere tray—the ones I have tried have all been very good—or make your own by buying separate components (the thermos and ice sphere molds to use are described in the section directly below). This thermos method takes up less space in your freezer than any product I've seen in stores, and it is more versatile when it comes to freezing objects inside the spheres. Commercially available clear ice molds tend to make somewhat smaller spheres than the thermos method, but on the plus side, they tend to make slightly rounder spheres. It's a trade-off.

Clear Ice Spheres: The Thermos Method

Clear ice sphere molds work the same way that clear ice cube molds do, with directional freezing. There is an insulated container (the thermos) and a sphere-shaped tray that sits partially inside it upside down, with a hole pointing down into the insulated reservoir. The water in the uninsulated ice sphere mold on top will freeze first from the top-down, pushing trapped air and impurities out of the hole in the bottom and into the thermos reservoir.

The setup I use at home is a stand-alone 2.5-inch (6.3 cm) ice sphere mold that sits on top of an insulated container, specifically a 10-ounce (300 ml) Thermos brand Funtainer food jar. The jar is short and wide, as it is meant to keep kids' lunches cold or hot. (One of my website readers, who probably has children, suggested this container.) This system has a total height of just under 6 inches (15 cm), which is shorter than any others I have seen.

Make Clear Ice Spheres in the Thermos System

1. Fill both the thermos and the ice sphere mold with water. As usual, you can use regular tap water if it tastes good enough, or use filtered water. No need to boil it.
2. Put your thumb over the hole in the ice ball mold to hold the water in and tip it upside down on top of the thermos as in figure 2.15. It will probably splash out a little as you drop it on top, but that's fine. The water already in the thermos will keep the water in the ice ball mold from draining out. It is also fine if the ice ball mold isn't 100 percent full, as water expands when it freezes and often fills the space.
3. Place the system in your freezer. In my freezer it takes exactly twenty-four hours for the ice ball part to freeze, leaving most of the water in the thermos unfrozen. Try to avoid leaving it in the freezer any longer than you need to.
4. Remove the ice ball mold from the top of the thermos; it will probably roll right off.

This insulated beverage container is the perfect size for a 2.5-inch (6.3 cm) ice sphere mold, but you don't need this exact model. The container can have a smaller circumference: You can use an insulated coffee travel mug or even a thick regular coffee mug. You might be able to use a foam beer koozie sleeve on a regular cup.

Ideally, you want the ice sphere mold to sit about halfway into the insulated mug (deeper into it is fine though) so that the directionally freezing ice pushes the water with trapped air out of the bottom hole into the water below. If your mug is too narrow, the hole in the bottom of the sphere mold can freeze closed, trapping the gassy water in the sphere portion. You want the water to freeze as slowly as possible, so remember to keep your freezer at a "warm" temperature as close to (but below) freezing as possible.

Figure 2.15. Ice spheres made in an insulated thermos. Fill the thermos and the sphere mold with water and rest the mold hole-side down on the thermos. Other shapes of insulated beverage containers can also be used.

The first couple times you use this setup, check the ice after eighteen or so hours to see if the sphere is already frozen solid. You want to avoid letting your thermos or container freeze too long—it might crack or shatter due to water expansion during freezing. If you take the system out of your freezer too early, you won't have a fully formed ice ball yet, but you won't have a cracked mug either. In my freezer at home, it takes twenty-four hours to freeze to the right level. Set a reminder alarm.

Troubleshooting the Thermos System

Is the ice cloudy at the bottom of the ice sphere near where the hole in the mold was? That probably means that the ice ball started freezing properly from the top down, but at some point the hole at the bottom froze closed. The remaining water in the mold was then forced to freeze from the outside in rather than pushing down through the bottom hole—and trapped air and impurities ended up freezing inside the sphere.

There are two reasons this usually happens. If your insulated coffee mug or thermos is relatively narrow compared with the width of your ice sphere mold, the cold air of the freezer surrounds more of the mold. Water inside the mold starts to freeze from the underside as well as from the top and the hole can freeze over. You can try to wrap a little scarf of some sort around the bottom half of the sphere mold to see if that will solve the problem.

Even if your insulated mug is nice and wide, the hole in the sphere mold can freeze closed if your freezer is very cold. See Appendix 1 for tips.

Another problem specific to this setup with the thermos and sphere molds is that the ice can come out more elongated (egg shaped) than perfectly round. This doesn't happen in

Figure 2.16. Edible flowers and other objects frozen inside clear ice spheres.

commercial clear sphere makers, as they tend to be anchored into the reservoir. Here, the sphere mold can separate slightly in the middle during freezing, and it freezes in this odd way.

To minimize the effect, first ensure that your freezer is warm and the thermos is protected from the freezer fan. Then put a little bit less water in the sphere mold and/or the thermos so that it doesn't expand and push the mold halves apart. Ideally, you'll find just the right amount of water to put into both your thermos and mold so that your spheres come out perfect.

Other Three-Dimensional Ice Sphere Molds

Most novelty ice cube molds are flat on the bottom, but there are a few molds in rounder shapes like diamonds, skulls, disco balls, and golf balls. These look particularly bad when they're cloudy, but they are fantastic when frozen clear using a directional freezing system.

Many of these molds fit inside the opening of a thermos just as sphere molds do. You might need to find a different sized insulated thermos though. The skulls in figure 2.17 are a little small and would have fit into a narrower-mouthed mug just as well. I found that the round-patterned mold came out clearest when I put it in the extra-wide Yeti brand thermos. This container is great because it insulates more of the top of the sphere, skull, or other molds you put inside it.

Clear Spheres in a Cooler

You don't need an insulated beverage thermos to make clear spheres in round or novelty shapes. Those are probably the most space- and time-efficient containers, but any of the directional

Figure 2.17. Novelty three-dimensional ice trays forced to freeze clear by using a thermos or other insulated beverage container.

Figure 2.18. Boulevardier on a clear ice skull.

Boulevardier on a Clear Ice Skull

1 oz (30 ml) bourbon
1 oz (30 ml) Campari
1 oz (30 ml) sweet vermouth

Stir all ingredients with ice and strain over a clear ice skull.

Garnish: orange twist.

freezing setups we have covered so far can be adapted to work for round shapes, too.

In a cooler, you can set your sphere mold (hole-side down) on a riser. The riser doesn't necessarily need to be a flat platform like we used to hold up perforated ice cube trays in figure 2.10. The platform can just be a pillar like a piece of pipe, or a narrow plastic container like a used yogurt container. The trick is to fill the water in the cooler to the right level, nearly halfway up the sphere mold, to encourage directional freezing without sealing the mold halves shut with water too high up its sides.

Or, instead of using a riser, you can press the ice sphere mold into one of the compartments of a 2-inch (5 cm) ice cube tray. Press the ice sphere mold down into the tray (hole-side down) just so that the mold is slightly submerged (rather than merely resting it on top) to encourage proper freezing.

You can set that ice cube tray on the bottom of the cooler or on a riser: if on a riser, you can harvest clear cubes out of the rest of the tray while also making your ice sphere. If you freeze two six-cube trays at a time, you can make about ten cubes and two spheres per batch.

Store Ice

After you empty an ice tray or cut up ice as we'll do in the next chapter, you'll likely have some wet ice. If you put this ice right back into your freezer, the cubes will stick to each other when the surface water freezes. It can be really hard to pry them apart after that.

Most of the time, I put my wet ice cubes into a plastic colander so that some water drips out of the bottom, or I place them in some other kind of plastic storage container or bowl. I put the ice back into the freezer in this bowl and set the timer on my phone for five or ten minutes. When the timer goes off, I pull out the plastic container and give it a little smack: the ice has started to stick together, and this breaks it up before it's too firmly stuck. I restart the timer, put the ice back in the freezer, and repeat this process a few times over twenty to thirty minutes until the ice is dry. Then I transfer it to a plastic bag or other container for longer-term storage.

Other ways to keep your wet pieces from sticking together (these are repeated in Appendix 1):

- Dry them off. Shake off the water and/or pat them dry with a towel or polishing cloth before putting them in the freezer. This won't prevent them from sticking together entirely, but it will help a lot.

- Don't let it touch. Separate your cubes on a plastic cutting board or tray when you put them back in the freezer. After a half hour or so, pry them off the board. Then you can store them together in a plastic bag or container.
- Use vodka. Some bar owners who make a lot of ice spray vodka on the ice before stacking pieces together. I'm not sure how well this works.
- Put cubes in individual plastic bags. Put each cube into a resealable plastic sandwich bag (don't seal them), and stack those up to dry. Wait thirty minutes or so for them to dry, and then you can combine them all in one larger container. Reuse the bags for the next batch!

After the ice has dried, or if it was never wet in the first place, you can combine it for storage. Clear ice doesn't become cloudy on the inside when stored—I've had people ask me that. The outside of the cube may get frosty (depending on your freezer), but the frost melts off quickly as soon as you take it out of the freezer again.

Ice shrinks in a typical freezer, however: it sublimates into a gas without passing through the liquid state, and thus your cubes get thinner over time. To prevent (or at least reduce) this, store your ice in a sealed container. Plastic food containers and resealable plastic bags work great.

The other great reason to seal up your ice is to prevent it from absorbing smells from food in your freezer and refrigerator. This happens surprisingly quickly—by the next morning any ice in your freezer's ice maker might smell like last night's garlic shrimp in the takeout box in the fridge. Transfer your smellier fridge and freezer foods to sealed containers, or store your ice in sealed containers—or better yet, do both!

Figure 3.1. Patterned ice: a group of clear cubes patterned with a meat tenderizer.

3

CUTTING AND SHAPING ICE

Cut Off the Cloudy Part of an Ice Block

If you have made a full block of ice in a cooler as in figure 2.4, or even a slab that you let freeze a little too long, you'll have a cloudy portion of ice on the bottom of the clear slab. Let's get rid of it.

Cloudy ice is easy to chop off a block because it is soft and full of trapped air. I find that a three-pronged ice pick is the ideal tool to use for this because you can scrape with it—sort of like an ice scraper for your car windshield. It takes me less than thirty seconds to scrape the cloudy part off an ice block. Figure 3.2 is the block from figure 2.4 being cleaned up so that only a clear slab will remain.

You can also use a (clean) chisel, hammer, or knife; anything flat, pointy, and heavy is good. You can run hot water over it to melt it, though that seems like a waste of time and water. However, if you want a smooth surface after scraping, you can run some lukewarm water over the top. Or follow the polishing tips in the section "Clean Up Ice to Look Nice" on page 68.

Cut Up a Slab of Ice

You have made a clear slab of ice. Now it's time to cut it up into cubes and other shapes. Many people are hesitant about making clear ice in a cooler because they think that cutting up slabs is going to be a lot of work, but trust me on this: cutting up an ice slab is the fun part! It may take a little practice, but the practice is *also* fun.

The thinner the slab of ice, the easier it is to cut it up into neat, square cubes. Most of the time I only fill my cooler about halfway with water and let it freeze for two days. It comes out about 2 inches (5 cm) thick. I find that this is the perfect thickness for cutting into big

Figure 3.2. Scrape off the cloudy bottom. Use an ice pick to remove the cloudy bottom of a block frozen in a cooler via directional freezing.

cubes, though you may want yours thicker or thinner. Many people like to produce the thickest slab possible so that they can harvest and cut up the most ice per batch.

How to Cut Up an Ice Slab

1. Allow your slab to temper (see below). Twenty minutes is probably long enough.
2. Set the slab on a wood or plastic cutting board rather than right against your countertop. The cutting board provides a little cushioning.
3. Score a line across the top of the slab with your saw, knife, or ice pick. For a slab of only a few inches (5 cm) thick, your line only needs to be about one-eighth of an inch (one-half cm) deep. For a thicker slab, you might want to go deeper.
4. If the slab is more than about 6 inches (15 cm) thick, turn the block and continue to score the line around the sides and also the bottom so that all the lines connect.
5. Set the knife into the groove of the scored line, or the ice pick in the middle of it, and give it a tap with something: a muddler, a hammer, the back of an ice pick, a block of wood, etc. Hopefully the block will split directly in half.
6. Repeat the process, splitting each piece of ice in the middle until you have a cube (or a longer spear). I find that it is easier to split ice from the middle each time rather than trying to cut off the side.

Facing top left, Figure 3.3. Ice slab ready to be cut.

Facing bottom left, Figure 3.5. Cutting a slab. Cut the slab in half each time.

Facing top right, Figure 3.4. Cutting up an ice slab. Score a line across the center of the slab and tap on it with a mallet.

Facing bottom right, Figure 3.6. Cutting a slab. Keep cutting the slab in half until your ice is the right size.

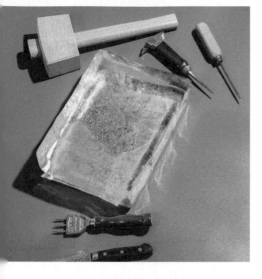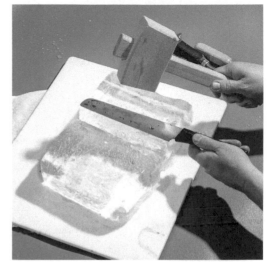

This process should produce straight-sided ice cubes. I'm often lazy, in a rush, and don't care about having straight edges on my ice cubes, so I don't often bother with all of the above. I'll pull the slab out of the cooler, use a three-pronged ice pick to score a line, and then plunge it into the center to break it in half. I can break down an entire thick slab into cubes in under a minute if I'm not fussy.

Ice-Cutting Tools

A Saw. When I first started cutting up ice slabs, I thought I needed a saw to cut all the way through the slabs, so I bought a wood saw that I only used for this purpose. I made sure to wipe it dry after every use to ensure it didn't rust. It worked okay, but I haven't used it in several years.

You do not need a saw for cutting up cube-sized ice made in a home freezer. There are specialty (usually Japanese) ice saws that have a round wooden handle and huge saw teeth. These cost hundreds of dollars and are designed for cutting much larger blocks of ice. You probably don't need one of these either, but I admit they look really awesome.

Knives. To cut up ice, you'll score a narrow line across the surface of an ice slab and then whack it to break it in half. A knife is great for this. I use a cheap bread knife, as it is long enough to reach across the width of the whole slab. Other people have used a pumpkin knife (looks like a bread knife but with bigger teeth), or a knife that looks like a cleaver. Most of these knives have teeth, but all of them have flat backs, which is the important part. You will hold the knife in the crevice and tap it on the flat top to help the slab break evenly.

Ice Picks. You don't strictly need a flat knife to break an ice slab in half. Most of the time I just use an ice pick, as I tend to have one handy. I find that the three-pronged pick is once again the most efficient tool for this and most ice-cutting and -shaping tasks. Tap the pick along the surface to score a shallow line across the top, set the pick in the center of the slab with its tongs in the crevice, then tap the pick with a mallet to split the slab in two.

Power Tools. Professional ice carvers use electric chain saws, drills, and other power tools to do their jobs. This is not a book on professional ice carving, so I haven't spent any time on these tools. You can try using a drill bit or electric screwdriver for any larger ice carving/designing projects. I have not attempted this, though I have purchased a wide drill bit to carve out a cavity in a big cube to make shot glasses. If you do this, be sure that every tool that touches the ice is cleaned for food safety.

Temper Ice

Tempering ice is just allowing it to warm up after you pull it out of the freezer. Ice tends to crack and split if it warms up quickly, especially when you're trying to cut it at the same time.

A fully tempered block will change from clear from your freezer to frosty and cloudy outside of the freezer and eventually back to clear and a little bit wet on its surface. Fully tempered ice is clear and glossy all the way through. You can let your blocks or slabs of ice fully temper before cutting them, but even a little bit of time outside the freezer helps.

I'm usually in a rush to carve up my ice, and sometimes I don't let it temper at all. Other times I forget about the ice tempering on my kitchen counter because I'll get busy doing something else. Luckily, clear slabs of ice melt really slowly, so even if I leave one out for nearly an hour I can still cut it up with most of the ice still frozen solid.

Clean Up Ice to Look Nice

Obsessed with smooth sides, straight lines, and perfect squares? This section is for you. After you've made your slabs and cut them into cubes, you still might want to flatten them further toward perfection. Even cubes, spheres, and other shapes made in trays often end up with a slightly cloudy or lumpy side you may want to remove. Rather than chopping the undesirable part off with an ice pick and leaving scrapes and scratches, you can simply melt off the imperfections.

Rubbing ice against the palm of your hand to smooth it out is surprisingly effective; however, for reasons including food-handling good practices and preventing cold hands, it is probably best to reach for a tool instead. A piece of conductive metal is ideal for smoothing the surfaces of ice.

Years ago I bought a thaw plate (also known as a defrosting tray or heat diffuser) for this task. These are just highly conductive metal plates used to speed up countertop meat defrosting. A thaw plate works well enough to smooth out ice—simply press the ice cube or slab against the metal, and it melts away. After smoothing out a few cubes the plate is less effective and you can "recharge" it after it gets cold by running it under hot water to warm it up.

Figure 3.7. Polishing ice. Use a thaw plate or the bottom of a pot to melt away imperfections.

Better yet, a tool you probably already own is a frying pan, spaghetti pot, or saucepan. The bottoms (particularly of fancy brands) of these pots and pans are usually made of conductive metal like copper. You'll want to make sure they're clean, of course.

There is no need to heat up the pans—they should work to smooth the sides of several cubes of ice at room temperature. Just press your cube or slab against the bottom of the pot. If I'm smoothing out a lot of ice, I'll set a spaghetti pot upside down in the sink and occasionally turn on the hot water so that it runs over the bottom surface of the pot to "recharge."

Of course, any other piece of flat metal should work as long as you heat it up with hot water. Look around your kitchen for anything like the side of a cleaver, bottom of a baking pan, and the like.

How to Hold an Ice Pick

An ice pick is useful for scraping down, splitting up, shaping, and refining ice. The way you hold an ice pick is important both for accuracy and to prevent injury, but it's not very intuitive. It seems that you *should* grab the end of the handle as far away from the pointy end as possible, but that's the more dangerous grip.

Hold your pick like you're holding a ski pole or walking stick, with the pointy side down. Slide your hand down the shaft of the ice pick nearly all the way, so that just a little bit of the pointy end sticks out below the pinky finger side of your fist. You should be gripping the middle of the handle rather than the top.

This grip gives you more control with your hand closer to the point of impact on the ice. Also, if your pick hand slips off the ice or your aim is bad, you're less likely to jab your other hand—or your knee, or the kitchen counter. If you do slip, only the bit of the pick sticking out a centimeter or two can plunge into you rather than the entire length of the pick.

As you build confidence in using an ice pick, you may be able to give yourself increased room by moving your grip up and allowing more of the pick to protrude.

For more delicate refining of smaller shapes like ice spheres, hold the pick like a pencil between your thumb, pointer, and middle finger in a "tripod grip." This is great for tapping away at the surface of ice rather than hacking away at it to break off larger chunks.

Carve Ice Spheres

This book presents several ways to make ice spheres: in molds, with an ice ball press (in the next section), or by hand carving them as in this section. We'll start with a big cube of ice and whittle it down into a sphere.

Chip-Away Method

This way is the safest way for beginners or people who never progress past the beginner level (like me). You can use a knife or an ice pick to make a sphere without too much risk of injury.

Hold a large cube (ideally about 3 inches, or 7.5 cm per side) of ice with your nondominant hand (your left hand if you are right-handed) in a cloth or towel. A glass polishing cloth works well because it doesn't have the longer fibers that can stick to the surface of the ice like a dish towel does. You can hold the ice without a cloth in your bare hand if you want, but

I find the cloth to be more secure. It also makes it easier to see what part of the ice to chip away next.

If using a knife, I use a big (and inexpensive) all-purpose chef's knife. Hold the ice lightly in your nondominant hand with your polishing cloth or towel and tap on the edges of the ice with the blade of the knife. Hold the knife by the handle as you would a sword and hit down on the edges of the cube like you're trying to chop it in half.

There is no need to hit the ice hard; you're just tapping it to break off bits from the edges. This process is often quite messy, with small bits of ice splattering everywhere and getting your kitchen wet. It's a great activity to practice outdoors in the summer.

The strategy is to hack away at the twelve edges of the big ice cube in your hand until the sphere reveals itself. Hit along each edge, turn the cube in your nondominant hand, and work on the next edge. For safety, the hand holding the ice should always be on the opposite side of the cube from the blade or ice pick.

If using an ice pick instead of a knife, a three-pronged pick works best for amateurs (like me) rather than a single-prong pick that requires more accuracy. Hold the ice pick more like you would hold a pencil: with your thumb, pointer finger, and middle finger gripping the metal that the points stick out of (known as a "tripod grip"). The handle of the ice pick should rest against the fleshy triangle made between your thumb and pointer finger. With the ice pick, tap the ice cube on the edges in the same places you would with a knife as described above.

After you've cut down all the edges, you should have something close to an ice sphere. It should be pretty obvious where you'll need to keep chipping ice off to get it just right.

Figure 3.8. Carving ice spheres. Chipping away at a cube to make a sphere.

Figure 3.9. Carving ice spheres. Carving a sphere like peeling an apple with a knife.

Peeling Method

Another way to carve a complete ice sphere—or to just clean up the remaining edges of one made using the previous technique—is to use a peeling motion. This requires practice, caution, and a smaller, sharper knife. It should only be done by those who have some basic knife skills (not me). As a general rule, it is bad form to bring the sharp edge of a knife toward your body, but this carving method requires you to do just that.

Think of peeling an apple with a knife like they do in old movies: hold the knife flat against the surface of the cube with the blade facing you and scrape off the surface ice as you bring the knife toward you. Ensure that your nondominant hand is on the opposite side of the ice from the knife, so that if the knife slips you will not cut yourself. Additionally, be extremely cautious of your thumb in your knife-holding hand: keep your thumb close to the blade—or on top of it rather than in front of it.

The peeling method can be used to clean up ice started in the chip-away method, but experienced ice handlers use this method to carve a cube into a sphere entirely. You may wish to watch a few videos of pros doing this to decide if you want to try it. (I choose not to.) You might also consider using cut-resistant gloves: these come in different materials with different levels of protection, from ones that can't take a direct hit from a knife to others made of chainmail for shucking oysters.

Polishing the Sphere

By using a polishing cloth, it is possible, if not fast, to rub the outsides of a cube down into an ice sphere shape. I once saw a Scottish barback polish a cube down into a sphere like he was polishing a bowling ball, but I haven't been able to replicate that. Instead, using the bottom of a spaghetti pot and a cloth, I've melted the outsides of a cube down to a lumpy sphere. I bet you can make a better one than I can.

Ice Ball Press

An ice ball press is an extremely heavy mold made from a conductive metal. By weight and heat transfer alone, it will melt a chunk of ice into a sphere in under a minute—no power

Figure 3.10. Ice ball press. The metal press has a sphere shape in the middle.

Figure 3.11. Ice ball press. With only pressure and heat transfer, the press makes beautiful spheres.

needed. The ice spheres that come out of an ice ball press look absolutely perfect, better than any other type of mold or hand-carving. But ice ball presses are relatively expensive, and you still must make the big clear ice cubes to put inside them in the first place.

As the cube placed into the press melts, the metal is drained of its heat and becomes colder. Depending on the size of the press and the ambient temperature in the room, you can melt just one or a few ice spheres inside a press before you need to warm the metal up again. You only need to run some warm water over it to do so. Bars that make a lot of ice spheres at one time with an ice ball press will often rest the press in a tray of hot running water so that it's constantly warming.

These presses come in different sizes. Cocktail Kingdom sells them in 70mm (2.75 inches) and 55mm (2 inches). Lately, other companies have been making them in other shapes like hearts and stars and golf balls. Still other companies make inserts that will press a logo onto the spheres made inside them. Search for "custom" or "personalized" "ice ball press."

Carving an Ice Diamond

This advanced-level trick begins with a cube and ends with diamond-shaped ice. It is easiest to start with a cube shape so you can follow along with the cuts below. For other projects, the chipping, scoring, and shaping of ice can be accomplished with a dullish knife, but for carving ice diamonds a sharper paring knife is ideal. As you grow more comfortable, you might decide to switch to a larger knife later on.

There are plenty of videos online on how to carve ice diamonds, though these tend to go in a slightly different order than the instructions below, which come from this book's photographer Allison Webber. I followed Allison's directions and made a passable diamond on the first try!

Facing top left, Figure 3.12. Carving an ice diamond. Cut off the bottom two-thirds of each side edge at an angle.

Facing top right, Figure 3.13. Carving an ice diamond. Cut the four remaining sides of the cube.

Facing bottom left, Figure 3.14. Carving an ice diamond. Turn the cube over and match the first cuts but cut straight down.

Facing bottom right, Figure 3.15. Carving an ice diamond. Make the bevels by matching the second cuts and cutting at an angle.

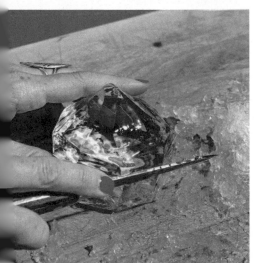

Carve an Ice Diamond

1. Place the cube on a cutting board or other nonslip surface. Be sure to hold the cube steady while making cuts.

2. Place the blade of the knife on one edge of the cube, about a third of the way down from the top. Cut down at an angle (roughly 45 degrees) in toward the bottom center of the cube where the long pointy end of the diamond will be. You probably won't be able to cut all the way to the center at that angle in one slice, which is fine. You can shave it down to refine it later.

3. Repeat the first cut for all four side edges.

4. On one of the flat sides of the cube (the only remaining flat part will be at the top), place your blade a third of the way from the top, connecting the imaginary line between two edge cuts you just made. Cut downward at the same angle toward the bottom center point.

5. Repeat this second cut on all four sides.

6. After these first eight cuts, you'll have the stubby bottom half of a diamond. You can refine those cuts into a sharper diamond point now or later. Flip the ice cube over.

7. You're now working on the wide, flatter top part of the diamond. That flat top is on the bottom against your cutting board. On the bottom uncut portion of the cube, align your knife to match the very first cut you made on an edge. But instead of cutting down inward at an angle, make this cut straight down.

8. Repeat this cut on all four edges.

9. Now match the second set of cuts: start at the bottom sides and cut down—at an angle again rather than straight down—to make bevels.

10. Repeat for all four sides.

11. You'll now have a stubby diamond. Go over the previous cuts to sharpen it up and make it thinner if desired. Fix yourself a drink when you're finished!

Figure 3.16. Revolver on an ice diamond.

Revolver on an Ice Diamond

2 oz (60 ml) bourbon
.5 oz (15 ml) Tia Maria coffee liqueur
2 dashes orange bitters

Stir all ingredients over ice and strain over an ice diamond.

Garnish: orange twist.

Branded and Patterned Ice

After all that effort to make perfectly clear ice slabs, spears, cubes, spheres, and diamonds, now we can slap a logo or pattern on top of it. At top cocktail bars and fancy hotel bars around the world, big cubes are often bedazzled with logos drilled deep into the ice. Those are beyond our abilities to make at home because they're usually carved by a CNC (computer numerical control) machine. This machine is a computer-controlled drill bit that can move in all directions above a sculpture-size ice block like a three-dimensional printer, except that it carves things out of the block rather than builds them up.

The CNC machine is used to carve logos, names, and designs into the flat face of ice sculptures for weddings and other events. (I always wondered how the ice carvers were able to make the logos so perfect; it turns out that a computer should get the credit.) For ice *cubes* with logos seen at cocktail bars, the CNC machine drills the logos onto the surface of a large slab of ice, and then afterward the slab is cut up into individual cubes, usually manually with a bandsaw.

Rather than building a home CNC machine, we'll investigate ways to put patterns, monograms, and custom logos onto your cubes in a more manageable way.

Branded Ice

Here's an easy project: purchase an ice brand and press it against your ice. Done!

The least expensive ice stamps are brass wax seal stamps, the kind used to press an imprint into a wax seal on a wedding invitation. You can find these online or in craft/paper/hobby stores for less than $10 each. They come ready-made in every letter and in a huge number of fancy fonts. There are also stamps with symbols like flowers, animals, and other shapes.

The brass is a great conductor and makes a nice (though shallow) imprint on your ice cubes. Many bars brand their ice cubes to order, but you can brand cubes in advance and return them to the freezer. They store well, especially in a sealed bag or container. (After a week or so, the outsides will sublimate, and your logo may become fainter.) I prefer to stamp cubes in advance and serve them right out of the freezer because the logos are easier to see when the cubes are frosty.

You can also have these stamps custom made for not very much money. Put your monogram or family crest on one—or a couple's initials for a wedding gift. (But also give them a clear ice cube mold so they can actually read the logo on the cube.)

You absolutely do not need a heated electric branding iron for this task, but you can find them for sale as well. These are often used in crafts and woodworking to burn designs into wood or leather, which requires more heat than pressing a shape into ice. Some bars do, however, use these electric versions because they can also burn logos into citrus peels for garnishes.

Barbecue meat brands tend to be much larger than the wax seal stamps mentioned above, and they also come premade in every letter, symbol, and font. Their disadvantage is that they are not usually made from highly conductive material like brass, so you will probably have to heat them up before use. I have done so by resting them in hot or boiling water before pressing them into ice cubes. These stamps can make deeper, longer-lasting logos on big cubes.

Top left, **Figure 3.17.** Tools for branding ice cubes: woodburning tool, wax seal stamps, steak branding iron.

Top right, **Figure 3.18.** Branded ice spear.

Left, **Figure 3.19.** Old-Fashioned on a branded ice cube.

Old-Fashioned on a Branded Ice Cube

2 oz (60 ml) bourbon
1 tsp or cube granulated sugar
3 dashes Angostura bitters
1 splash water

Brand your ice cube. Muddle water, sugar, and bitters in a stirring pitcher until dissolved. Add bourbon and ice and stir. Strain over branded ice cube in an Old-Fashioned glass.

Garnish: orange twist.

Patterned Ice

Much like pressing a logo into a cube, you can press a pattern into one. The trick is to find a good pattern maker. Some wonderful genius (not me) came up with the idea of using a meat tenderizer. It creates spectacular deep, spiked patterns in ice as in figure 3.1, and most tenderizers are just about the right size for a 2-inch (5 cm) cube.

The ideal material for a pattern maker is something highly conductive with patterns protruding about half a centimeter (or an eighth of an inch). You might also look into potato mashers and other objects in metal webbing or grid patterns. I have tried several different patterned objects I had lying around the house, like a drying rack for baked goods that worked okay.

But the real innovation in patterned ice is just beginning. A couple of ice tray manufacturers have released metal plates specifically for making patterns on ice cubes and spears

Figure 3.20. Patterned ice tools: meat tenderizers, vegetable masher, cooling rack.

(see Appendix 2: "Buying Guide"). They look fantastic, but they are quite expensive for now.

If you're going to pattern a bunch of cubes at once, you might want to heat up your meat tenderizer or other pattern maker. When I patterned all the ice for figure 3.1, I set the meat tenderizer in a shallow pan with a little water in it atop a stove burner. I pressed the ice down onto the meat tenderizer on each side of the cube. The heated pan of water kept the tenderizer warm.

Avoid using a culinary blowtorch or other open flame on any metal not meant for it, as the high heat can degrade cheap metal coating and potentially leak toxins onto your ice. Hot or boiling water is safer.

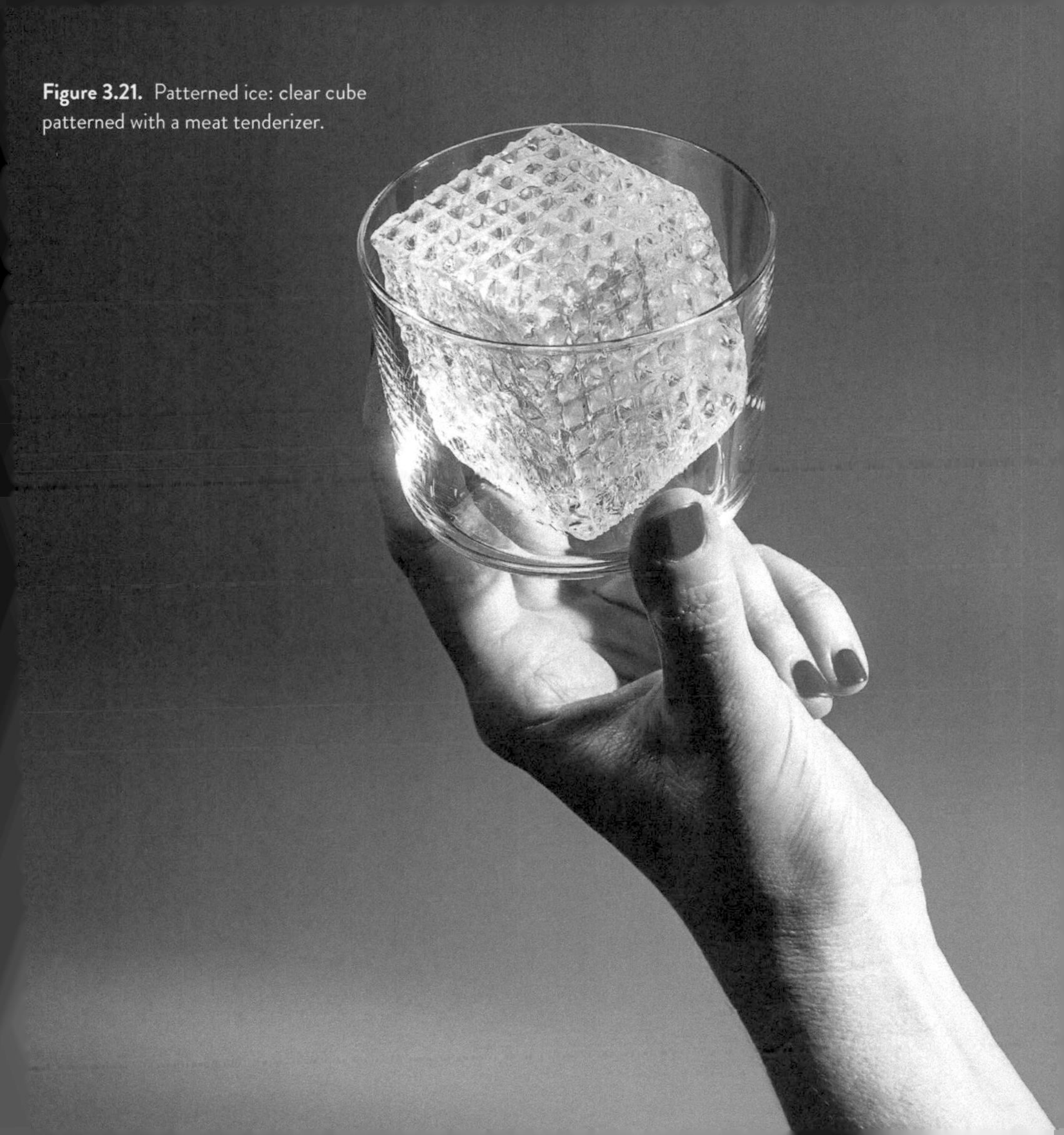

Figure 3.21. Patterned ice: clear cube patterned with a meat tenderizer.

Figure 4.1. Mint leaf in a clear cube. Fresh herbs like mint, sage, basil, and rosemary freeze beautifully inside clear ice.

4

FREEZING OBJECTS IN ICE

Freeze a Bottle Inside an Ice Block

To freeze a bottle inside a clear block or slab of ice, you'll just need a cooler and possibly some risers on which to sit your bottle. I have frozen some bottles completely inside ice blocks for table displays at events, and others with their tops exposed so you can actually open them up and pour out the liquids. Either way, bottles frozen inside blocks are huge crowd pleasers at parties. Using these tips, you can make these frozen bottled cocktails for other people or for yourself at home.

For short bottles, you can stand them upright in a cooler. You may choose to sit the bottle on a short clear riser if you want to keep it elevated above the cloudy bottom of the block. I used fake plastic ice cubes as a riser beneath the bottle in figure 2.12.

Short bottles can also be laid horizontally inside a cooler on a riser with the label facing up, as in the photo in figure 4.3. This way you only need to fill the cooler to just above the top of the bottle rather than the entire cooler. You won't need to freeze the block solid; you can remove the slab containing the bottle after a couple of days.

For taller bottles, you can lay the bottle diagonally in the cooler from the bottom of one corner to the top of the opposite side. (This is not pictured in this book.) After the block is fully frozen, you can remove it from the cooler and use an ice pick to hack off some of the excess surrounding ice, leaving a diamond-encrusted-looking ice shell around your bottle.

Pictured in figure 4.2, a way to gain some extra height around the bottle neck is to use a plastic pitcher inside the cooler. This also makes a clear sleeve around the bottle. Fill the cooler as well as the pitcher with water to encourage directional freezing. In the picture I didn't use a riser for the tall gin bottle, as the resulting ice, shown in figure 4.4, is stunning even without it.

Now you just need to decide what cocktail to freeze. I chose a Martini.

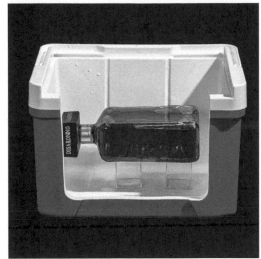

Top left, **Figure 4.2.** Freeze a bottle in a block. The pitcher makes a sleeve around the bottle, while the insulated cooler encourages directional freezing.

Top right, **Figure 4.3.** Freeze a bottle in a block. Rest a bottle on its side on a clear riser to make a display.

Left, **Figure 4.4.** Freezer Martini and ice cubes with olives frozen inside.

Freezer Cocktail Tips

- Don't freeze beer, wine, or nonalcoholic liquids. They will freeze solid and likely crack your bottle.
- Freezer cocktails should be high in alcohol content. For the best flavor, these cocktails should not contain juice and of course no carbonated liquids. Cocktails that are stirred rather than shaken usually fit this category.
- Common freezer cocktails include the Martini, Manhattan, Negroni, Sazerac, and Old-Fashioned.
- Predilute freezer cocktails. Most stirred cocktails are diluted by around 15 to 25 percent. Calculate your ideal level of dilution: measure out a cocktail such as a 3-ounce Manhattan, stir it with ice until it's just the way you like it, then strain it and measure the final volume. Then extrapolate that percent dilution to your freezer version. Frozen cocktails taste weaker than they are because the low temperature numbs the palate. If you don't predilute your cocktails you or your guests are likely to feel the effects too strongly!

Freezer Martini (makes 750 ml)

12 oz (350 ml) gin
6 oz (175 ml) of dry vermouth
10 dashes orange bitters
4 oz (120 ml) water

Combine all ingredients in a 750ml bottle and store in freezer inside an ice block.

Garnish: olives or a lemon twist.

Figure 4.5. Freezer Martini and ice sphere with olive frozen inside.

Other Objects in an Ice Block

The insulated cooler is well suited for freezing large objects that don't fit inside cubes or spheres. You can make punchbowl ice, displays, and centerpieces with ice blocks.

For objects that float, the challenge is keeping them submerged in water while freezing. Flowers, for example, are extremely buoyant and hard to keep underwater. (Consider fixing them in the cooler held upside down by their stems instead.) Other items that float flat on the surface, like some citrus peels and herbs, end up being covered by a thin layer of ice as it freezes. This might be good enough for some displays.

Help submerge floating objects with rods, tape, or twine laid across the surface of a cooler. Once the surface of the ice freezes over, the ice holds objects in place and you can remove any supports. You might also anchor objects from below like a buoy with fishing line tied to something heavy that sits on the bottom of the cooler.

Objects that sink in water are their own challenge, but you can hang them from the cooler handle or from something like a dowel or strong tape laid across the cooler's surface. In figure 4.6, I hung the shark from a string attached to the cooler handle, with just the fin of the shark sticking above the surface of the water.

For the strip of citrus peel spelling "HELLO" in figure 4.8, first I made the orange peel and punched out the letters using aspic cutters. Then I draped the peel like a banner across the top of an open insulated cooler, with the ends of the peel taped to the sides of the cooler out of the water with the letters facing downward. I filled the cooler all the way up with water so that the peel was submerged. This made a nice punch bowl ice block for my Mojito after I removed the slab and flipped it over.

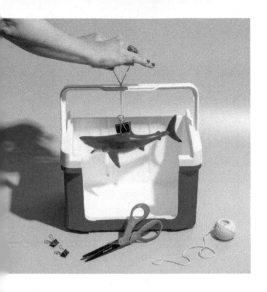

Top left, **Figure 4.6.** Freezing objects in a clear ice slab or block: hang objects that will sink by string from the cooler handle.

Top right, **Figure 4.7.** Freezing objects in a clear ice slab or block: frozen shark in a clear ice slab.

Left, **Figure 4.8.** Freezing objects in a clear ice slab or block: a citrus peel that was hung from the top of an insulated cooler.

Other objects that sink can be set upon a clear riser or pillar like the Disaronno bottle in figure 2.12. For risers I have used clear plastic boxes, plastic ice cubes, and plastic containers. Allow the water to freeze to a level just below that of the object, remove the slab, and pull out the riser from the ice.

Food Safe Objects and Practices

The list of things I have frozen into ice blocks, cubes, spheres, and spears is long and stupid. It includes a pair of sunglasses, a fake severed hand, various office supplies, and a pile of loose change. None of those objects I would have served to someone in an actual cocktail. Keeping your ice food safe is an important topic that we'll discuss in this section.

There are two issues to consider: whether something you've frozen into an ice cube can be dangerous or poisonous on its own—and whether someone can be hurt by it.

Mojito Punch with a Big Ice Block

2 oz (60 ml) white rum
1 oz (30 ml) simple syrup
1 oz (30 ml) lime juice
2 oz (60 ml) soda water
1 sprig mint

Scale up to suit your punchbowl. Shake rum, syrup, lime, and a few mint leaves with ice. Strain into punchbowl and add soda water, ice block. (Traditionally served in a highball glass with ice and mint garnish.)

Garnish: mint ice cubes.

Figure 4.9. Mojito punch with a big HELLO ice block and mint cubes.

Speaking to the latter, both children and intoxicated adults will swallow anything within arm's reach, so strongly reconsider putting anything you don't want swallowed within arm's reach of children and drunk people. Examples include cherries and olives frozen into ice spheres: not a problem in regular drinks but a choking hazard when frozen. There are situations in which I'd serve inedible but food-safe objects frozen into ice cubes to my friends at my book club, for example, but never to strangers at a bar or a wild party. Know your audience.

Additionally, be cautious with inedible objects that are not intended for food use like plastic toys and decorations. They're probably not food safe and *especially* not alcohol safe. Alcohol can leech dyes or other chemicals out of plastics, for example. I freeze a lot of cheap plastic junk into ice cubes like Halloween spider rings and novelty eyeballs, but I don't put those cubes into drinks with alcohol.

Passion Project on an Edible Flower Ice Sphere

2 oz (60 ml) white rum
1 oz (30 ml) passionfruit syrup (Small Hand Foods brand recommended)
1 oz (30 ml) lime juice
1 oz (30 ml) ginger beer (optional)

Shake rum, syrup, and lime with ice and strain over an edible flower ice sphere. Top with ginger beer if desired.

Garnish: none.

Figure 4.10. Passion Project on an edible flower ice sphere.

Keep in mind that just because something is *natural* doesn't mean it is *edible*. A beautiful leaf you picked on a hike might look pretty when frozen into an ice cube, but it might also be poison oak. Stick to flowers and other plants you've verified as edible or safe. Some organic grocery stores sell packs of edible flowers, like the marigolds I froze into the drink in figure 4.10. Check my website CocktailSafe.org for lists of edible and inedible flowers as well as other information about food safety as it relates to cocktails.

Freeze Objects Inside Ice Cubes and Spheres

Everybody loves a groovy cube with something frozen inside it. I have seen people so thrilled with their ice cubes that they reused their cubes in multiple drinks, or took an extra ten minutes to finish a beverage because they wanted to experience licking the fruit inside the cube; they even asked for a plastic bag so that they could take the rest of their ice cube home with them. Give the people what they want, I say.

It is not particularly challenging to freeze objects inside ice cubes or spheres, but you will need to decide which type of ice tray—sphere or cube—you want to insert an object in for maximum impact. Most clear ice cube trays, including the ones we made by poking holes in a silicone tray, are open at the top. This makes them convenient for freezing objects that sink, as well as objects that fit diagonally across an individual cube compartment.

Sphere molds on the other hand are not usually open on top (though some commercial clear sphere molds do have a fill hole there). Objects that are buoyant may not freeze well in open-top *cube* trays, but in sphere trays the objects are trapped inside the mold and look great.

Figure 4.11. Centering objects in clear ice trays. Hang a cherry from its stem to put it in the middle of the sphere.

Figure 4.12. Strawberry in a clear cube made in a clear ice tray. A metal cocktail pick was used to hold the strawberry in the middle of the cube while freezing and can be removed easily.

One tip for small floating objects like cranberries: if you put just one of them into a sphere tray it may look odd and "stuck" to the inside edge of the finished sphere, but if you fill your sphere tray with five of them some will end up in the middle. Sphere trays are also better suited to slightly larger objects and ones that feather out, like edible flowers and herbs.

For objects that sink, when they rest on the bottom of either type of tray, they can sometimes plug the holes there and prevent the trapped air and impurities from escaping during directional freezing. Depending on what you're trying to freeze, you may be able to hang the object inside the tray (as long as it has a hole at the top). I have used binder clips to hold the stems of cherries (as in fig 4.11) and have also inserted a metal cocktail pick through strawberries and olives so that the fruit was right in the middle of the cube or sphere. The metal picks slide out quite easily after freezing, but they also look pretty cool if you leave them in.

Objects in Ice Tray Tips

- After placing the object to freeze inside an ice sphere mold and filling it with water, tap it or shake it to eliminate any air pockets.
- Juicy objects like thick lime wheels or cucumber slices can get squished as they freeze and push out some juice into the ice, with cloudy results. Keep your slices thin, and use drier produce.
- The objects that look particularly good are ones that fit nearly from end to end in an ice ball mold (like edible flowers or a basil or mint leaf), or cube tray (like a citrus peel). When touching both sides of the tray, objects hold their place instead of sliding to the bottom while freezing.

Ideas for Edible Things to Freeze Inside Ice Spheres, Spears, Cubes, and Slabs

For inspiration, just walk around the grocery store or the farmers market seeking small items that might be freezable. Consult books, websites, and social media sites for garnish tips: many of the citrus peel shapes, carved vegetables, and other garnishes made to accompany food or decorate drinks can also fit inside cubes.

Citrus Peels

Consider a cocktail's garnish, and put it inside a cube instead. Citrus peels are flat as well as easy to cut and twist into shapes. To peel citrus, you can use a knife, but a Y-peeler makes nice, consistent, wide peels. A Y-peeler is an inexpensive tool (shaped like the letter Y) that is surprisingly sharp and dangerous. Read the directions or look up a video on how to use them, or you will very likely peel some skin off a finger. The trick is to hold the peeler steady on the surface of the citrus with one hand and rotate the fruit with the other hand. Rotate the fruit, not the peeler!

Once you have cut your peels off the fruit, you can trim them into twists and knots, curls and strips, circles and squares, and other shapes using a paring or other knife.

Citrus and Vegetable Cutouts

Punch out shapes from citrus peels using aspic cutters—or, better yet, a "vegetable cutter set." Aspic cutters look like tiny cookie cutters, with one side slightly sharper than the other. They're designed to make shapes out of soft things, so you must press fairly hard on a citrus peel to punch it out fully. Aspic cutters come in all sorts of letters, animal and botanical shapes, holiday themes, and more.

Vegetable cutters are the same as aspic cutters but made for solid fruits and vegetables: so they come with plastic protectors that save your fingers. These are sometimes used for bento boxes and kids' lunches and typically come in shapes like hearts and stars. They make adorable garnishes.

Citrus and Vegetable Slices

Slice small limes or English cucumbers into wheels. Or start with a carrot slice and use the vegetable cutters mentioned above to make shapes. Remember that juice is the enemy of clear ice, so before freezing make thin slices and/or let them dry out a little after cutting. I use cucumber slice ice in spa-style cocktails with mint and lime.

Fresh Herbs

Mint, basil, dill, rosemary, cilantro, parsley; these look so good in ice because they're nonporous. Rosemary and mint sprigs also look great in long spears, as in figure 4.16. Dill, cilantro, and parsley are particularly great in sphere shapes since they spread out like a carnation. Mint and basil look great in any type of ice.

Berries and Cherries

Bright berries of all sorts freeze well, look great, and reinforce the refreshing nature of cocktails. Some of the most stunning cubes I've created are a single strawberry or cherry on a stem suspended in the middle of a cube or ice sphere. I haven't had success with the bright red maraschino cherries, though, because the sugary syrup they're stored in doesn't freeze. Fresh cherries seem to work best. For small berries like raspberries, blueberries, and cranberries, try putting several inside an ice sphere so that some will be stuck in the middle, not just at the outer edge.

Edible Flowers

See the section on food safety for more information, but I repeat: natural does not mean edible. Purchase or grow edible, nonpoisonous flowers to freeze into cubes. You might find edible flowers at the farmers market, and some natural-leaning grocery stores carry packaged edible flowers.

Tiki bars and Thai restaurants often serve food and drinks with purple and white orchid garnishes. These freeze really well inside large spheres and last for a long time both before and after freezing. The only challenging part is finding them for sale: many bars and restaurants buy them in their large food orders from purveyors that consumers do not have access to. There are also companies that specialize in mail-order edible flowers you can find online.

Inedible Things to Freeze Inside Ice Spheres, Spears, Cubes, and Slabs

Perhaps it is silly to separate edible and inedible items into different lists. If a person swallows a frozen strawberry, they are still going to have a negative digestive experience. However, it is important to reinforce that nonedible and non-food-safe items should be considered differently and used with discretion—sometimes only as decoration on the side rather than garnish in the glass.

Flag Cocktail / Hors d'Oeuvres Picks

Cut down the toothpick part with scissors so that the flag will fit inside your cube or sphere. These are ideal for occasions where there is a lot of flag waving, such as the Fourth of July, LGBT Pride, international sporting contests, and Eurovision.

Figure 4.13. Freezing objects in a clear ice tray. Make letters with aspic cutters or insert berries, edible flowers, and citrus shapes into the tray.

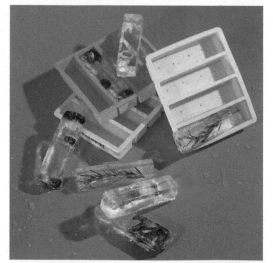

Top left, **Figure 4.14.** Edible objects in ice cubes including carrot spirals, mint, and citrus shapes and twists.

Top right, **Figure 4.15.** Gin Rickey over YES ice cubes.

Left, **Figure 4.16.** Clear ice spears made in a perforated tray with blueberries, citrus peels, rosemary, and mint.

Gin Rickey over YES Ice Cubes

2 oz (60 ml) gin
1 oz (30 ml) lime juice
4 oz (120 ml) soda water

Stack cubes in a highball glass. Add gin and lime then top with soda water.

Garnish: none.

Halloween Stuff

I go a little wild every Halloween. I have frozen a variety of masks into ice blocks, bloodshot eyeballs into spheres, a plastic severed hand in a punchbowl ice block, skeletons into spears, and all sorts of spiders and other plastic insects into cubes. Some of the best Halloween objects to freeze are plastic vampire fangs, as they're inexpensive, come in multiple colors, and you can fold them up and freeze them inside a sphere ice ball mold.

Holiday Ornaments and Home Decor

Small hanging ornaments come in all sorts of flowers and figures tied to specific holidays or just for general decoration. Check out housewares, holiday, and craft stores—they have everything from mini dollhouse items to plastic jewelry to freeze as well.

Figure 4.17. Inedible objects in clear ice: babies, flags, skeleton, spiders, dice, and an eyeball. See text about food safety.

Figure 4.18. I Like You. Printed on a shrink plastic sheet. Not edible.

Toys and Games

I buy a lot of freezable items from the gumball machines at gas stations and old malls. They're usually fifty cents or less and already fit into sphere-sized containers! Also, items related to gaming nights—poker chips, dice, board-game pieces, and the like—make great novelty cubes.

Party Swag

Various baby shower party games involve putting a tiny plastic baby into an ice cube or other food item. Put them into clear ice cubes instead. Other party gift bag items can make fun novelty ice displays as well—check out your local party store for inspiration.

Bachelorette parties ("hen parties" to you Brits) often come with a lot of items in the shape of a male member. Though many of these would look hilarious frozen into ice cubes, these parties often get a little wild, and perhaps it is best to avoid the combination of alcohol and a choking hazard.

Drink Inside an Ice Sphere

This trick was pioneered at the molecular mixology cocktail bar The Aviary in Chicago. Let an ice sphere freeze for only a few hours to create an outer shell, puncture it to release the interior water, and then refill the sphere with a cocktail.

This works best with stirred cocktails like the Martini, Manhattan, Negroni, and Old-Fashioned that freeze well and don't require fresh citrus because you'll want to prefreeze the cocktail before filling the sphere with it.

Make a Drink Inside an Ice Sphere

1. Fill an ice sphere mold with water and place it in the freezer. (Actually, put several in the freezer so you can test to see when they're ready.) Do not use a directional freezing system: in this case we *want* the ice to freeze from the outside in. You can even use a water balloon for this; just be sure to flush out the powdery stuff inside it before filling it in the sink.

2. Mix your cocktail, and store it in the freezer. You want it to get as cold as ice so that it doesn't melt the ice shell later.

3. After about three hours of freezing, test a sphere to see if the ice is thick enough to poke a hole through the surface without smashing the whole sphere. If not, wait another hour and try again. Too thick of an ice shell means it will be harder to poke the hole; too thin means your shell is more likely to crack during handling.

4. To poke the hole through the ice sphere, The Aviary staff use a drill with a tiny drill bit. I have used a narrow metal straw, a pointy ice pick, a knife, and the end of a metal cocktail pick heated up in hot water to melt the hole.

5. Drain the water. You can shake the water out (you may need to insert a straw or eye dropper or something else inside to encourage it). Bars making a lot of these often use a syringe to suck out the water. They also use the syringe to reinsert the cocktail.

6. Fill the ice shell with your cocktail. Use a syringe or a pipette or tiny funnel.

7. Serve. Set the cocktail in an Old-Fashioned glass. To smash the cocktail open, use the back of a small spoon, a mini muddler, or a mini hammer.

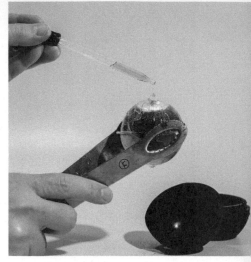

Top left, **Figure 4.19.** Cocktail inside an ice sphere: Freeze water in a sphere mold for a few hours, then empty out the water.

Top right, **Figure 4.20.** Cocktail inside an ice sphere: Refill the ice sphere shell with a frozen cocktail.

Left, **Figure 4.21.** Manhattan inside an ice sphere with hammer to crack it open.

Manhattan inside an Ice Sphere

2 oz (60 ml) rye whiskey
1 oz (30 ml) sweet vermouth
2 dashes Angostura bitters

Stir all ingredients with ice, strain into a new glass, then move to the freezer while preparing the ice sphere. Insert into the ice sphere in an Old-Fashioned glass. (Traditionally served in a cocktail glass and garnished with a maraschino cherry.)

Colored and Flavored Ice

There is a lot of overlap between colored and flavored ice. Sometimes you only want your ice to look different; other times you want it to taste different too. In either case, directional freezing is not your friend.

Remember, trapped air and impurities are pushed away from the point of freezing, and the last part of the water to freeze is where those impurities end up. Color and flavor compounds are impurities, so using colored water in a directional freezing system means that most of the color ends up in the cloudy section of the ice.

When working with colored and flavored ice, choose smaller ice cube trays, as they freeze quickly from all sides. Even larger sphere molds and 2-inch cube molds can freeze unevenly when colored and flavored liquids are inside.

Not all liquids freeze well, or at all. Alcohol freezes at much lower temperatures than water, so only bars with professional equipment can usually accomplish freezing alcohol or liqueurs. Sugar also makes for slushy, sticky ice. The ideal liquids for freezing don't have alcohol, many solids, or sweeteners. They're just watery and colored or flavored like tea. But don't let that stop you from experimenting.

Color-Changing Ice

In recent years blue color-changing drinks have become popular in cocktail bars, and it is all thanks to butterfly pea flowers. These flowers (the product is often marketed as "butterfly pea flower tea," but it is just dried flowers in tea bags) dye water and other liquids a bright blue color. Then when acidic ingredients like citrus juice are added to it, the color changes first to a deep purple then to a bright pink color. As these flowers don't carry a lot of flavor (just a little musty/earthy flavor if you use a lot), you can brighten a lot of different cocktails with this one natural food color.

For your drink to start out blue, the liquid into which you infuse butterfly pea flower must be pH neutral like water, ice, or simple syrup. Some alcoholic spirits will retain the blue color when infused with it, but many spirits are slightly acidic and will turn purple instantly. This looks great too.

However, when assembling your cocktail, any neutral-pH blue liquid will almost certainly mix with an acidic ingredient like a liqueur, citrus juice, or carbonated water. Most of the time your cocktail will remain pink after mixing.

To show off the color-changing property of butterfly pea flower, infuse it into hot water, then freeze the blue water into ice. It doesn't take many flowers to make your water a pleasant blue. Avoid overdoing it with the flowers, as the wilted vegetal note can be a distracting flavor.

When you add your ice to the rest of your cocktail, the ice should remain blue until it starts melting, slowly turning the drink purple-pink. The impact will be greatest in drinks that are mostly clear or light in color. Some options are the Gin and Tonic, Gin Rickey, Mojito, Gin/Vodka Collins/Soda, Margarita, and Moscow Mule.

Blue and Purple Ice

Butterfly pea flower isn't the only plant that does this color-changing trick. You may have seen it in grade-school science class when using juiced red cabbage as a pH indicator. The cabbage works the same way in cocktails but doesn't taste so great! Blueberries are a little better.

All these plants are blue and purple due to anthocyanin pigments. Other plants that have anthocyanins include purple corn (the type used to make chicha morada), Norton grapes, eggplant skins, blueberries, black raspberries, raspberries, Concord grapes, acai, plums, ube, and blackcurrants. Perhaps you can do something magical with them as well.

In each section of colored and flavored ice below, I'll name some freeze-dried powders you can buy to turn your ice that color. There are companies that now specialize in making powders primarily for food coloring, available in consumer-sized quantities. In the blue range, blue spirulina is now also available as a colorant, and when acids are added it turns a wild milky turquoise color. Other blue or purple powders include ebony carrot and sapphire wolfberry.

Figure 4.22. Color-changing Gin and Tonic made with butterfly pea flower ice.

Gin and Tonic with Color-Changing Ice

2 oz (60 ml) gin
.5 oz (15 ml) lime juice
3–4 oz (90–120 ml) tonic water

Make ice by freezing butterfly pea flower water. Add all ingredients to a wine goblet or other glass over ice.

Garnish: none.

Red, Yellow, and Green Ice

To make red ice, you can use dehydrated beet powder (or freeze beet juice), hibiscus leaves, or rooibos tea. I have had less success with *juicing* strawberries, raspberries, and cranberries, but if you freeze these berries into cubes whole, you'll have red ice anyway. In commercially available dehydrated powders, dragon fruit produces a bright red-purple color and is made by many brands.

For flavored ice, red and purple juices include cranberry, cherry, pomegranate, and purple grape juice. Keep in mind that juices with a lot of sugar in them end up making goopy rather than hard ice. That's another reason to make them in smaller trays: they come out less slushy that way.

Angostura and Peychaud's bitters, used in many classic cocktails, dye liquids a reddish color and can have a big flavor impact. If you like your Gin and Tonic with bitters, try putting the bitters into the ice so that the drink changes flavor over time.

In the yellows, saffron is a famously effective (and famously expensive, but you don't need much) yellow colorant that doesn't impart too much flavor to ice/drinks. It gives more of a tannic sensation. Other spices that may work include annatto (with a slightly nutty/peppery flavor), turmeric (earthy and a little spicy), and mustard seed. Due to their strong flavors, you may need some advanced mixology skills to use them properly. In dehydrated colored powders, I have spotted something called "yellow goldenberry" for sale. Yellowish juices for flavored ice include pineapple juice, orange juice, and mango.

Fresh herbs like mint and basil look great frozen into ice cubes, but, like most green plants, once you pulverize or juice them, they start turning yellow-brown really fast. If I want green ice, I'm more likely to freeze a whole mint leaf, a lime peel, or a cucumber slice into a cube rather than use the juice.

That said, green tea and matcha teas stay green in water, and you can find dehydrated green spirulina, pandan leaf, wheatgrass, and other powders that retain their color after you rehydrate them. Again, you may be working with some unwanted flavors, so it is best to put these vegetable/grassy powders into flavorful/citrusy/creamy cocktails that will minimize their impact.

Black Ice

Activated charcoal has been infused into lemonade, ice cream, toothpaste, and a lot of "detoxifying" foods. It doesn't really work, but it does turn them an extremely pretty goth black color. In ice, the charcoal tends to settle out, and the ice looks gray at best. However, there is another reason not to use activated charcoal in drinks: it can disable any necessary medications consumed within a few hours before or after eating it. I advise people not to use this

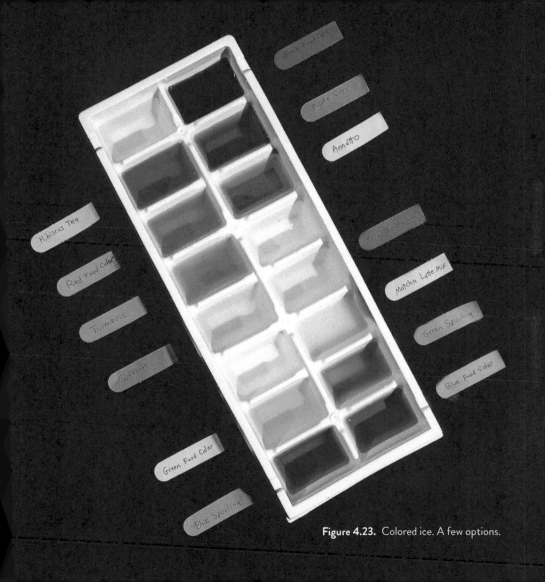

Figure 4.23. Colored ice. A few options.

Figure 4.24. Colored ice in water balloons.

ingredient on cocktail menus without a warning; you don't want to interfere with anyone's medication.

Other black items that can be used to dye cocktail ingredients are black sesame seeds and squid or cuttlefish ink. Black sesame also tends toward a gray color when pulverized, and the inks taste and smell slightly fishy and salty, but some bartenders have used them to great effect.

For black and almost every other color of ice, you can always reach for artificial food coloring; the kind used in Easter egg dyes.

White Ice

White ice isn't particularly exciting on its own, but it helps other colors pop, as it does with the Queens Park Swizzle (fig 4.25). Milky-colored liquids like coconut and aloe waters make whitish-colored ice. I haven't tried freezing rice or oat milk into ice cubes, but if you do, let me know if it's any good!.

Probably the easiest way to make very white ice is to ensure it is full of air because, as we know, that's what makes ice cloudy. Use carbonated water in your ice trays to maximize air in cubes, or make white ice by crushing it.

Things to Do with Colored and Flavored Ice

The easiest way to use flavored ice is to take one of the ingredients from a cocktail and put it in the cube instead of (or in addition to) it in the liquid. So, your Cosmopolitan could be made with just lemon vodka, triple sec, and lime but served on a cranberry ice sphere.

The second easiest way is to consider a flavored version of a classic drink (Pomegranate Mojito, Strawberry Daiquiri, Pineapple Margarita) and put that added flavor into the cubes, leaving the classic recipe alone.

Other Ideas:

- A Mimosa with orange juice ice spheres (fig. 4.26)
- Coffee cubes in a White Russian or Espresso Martini
- Hibiscus tea cubes in rum drinks like the Mojito and the Dark and Stormy
- Pineapple juice cobblestones in a Chartreuse Swizzle or Bahama Mama
- Coconut water cubes in a Painkiller or a Swamp Water (fig. 4.27)
- A Bloody Mary with celery juice ice spears

Queens Park Swizzle

3 oz (90 ml) white rum
1 oz (30 ml) simple syrup
1 oz (30 ml) lime juice
3–4 dashes Angostura bitters
mint leaves

In a highball glass, lightly muddle about six mint leaves in the bottom of the glass. Add liquid ingredients, then fill the glass with crushed or cracked ice. Stir or swizzle to make the glass frosty, then top with more crushed ice. Dash bitters on top of ice, then garnish.

Garnish: mint sprig.

Figure 4.25. Use crushed ice to make the red and green elements of the Queens Park Swizzle stand out.

Other Ice Shapes

In this section we'll look into smaller types of ice and how to manipulate those too.

Cobbler, Cracked, or Crushed Ice

These types of ice are usually broken down from larger pieces, so they're solid but smaller than cubes. Cobbler ice in the olden days consisted of hailstone- or pebble-sized pieces of ice that broke off from a large block when it was being cut up. Cracked or crushed ice is now broken down from cubes, usually with machines. Crushed ice is usually the smallest. These types of ice are used in cocktail categories of swizzles, cobblers, and juleps.

There are several ways to make this type of ice at home. Old-timey hand-crank ice crushers (fig 4.32) have a handle and ice-crushing claws inside them. There are also electric versions of these. Or you can throw your ice cubes into a blender and use that to pulverize your cubes down to your desired size.

Mimosa with Orange Juice Ice Spheres

3 oz (90 ml) sparkling wine
Orange Juice Ice Balls

Freeze orange juice into cobblestone or other small ice cube shapes overnight. (Cranberry ice also works great.) Fill a flute about three-quarters full with sparkling wine, and drop orange juice ice balls into the drink.

Garnish: none.

Figure 4.26. Mimosa with orange juice ice spheres.

Figure 4.27. Swamp Water with coconut water ice cubes.

Swamp Water with Coconut Water Ice Cubes

1.5 oz (45 ml) Green Chartreuse
4 oz (120 ml) pineapple juice
.5 oz (15 ml) lime juice

Add all ingredients to a glass filled with coconut water ice cubes.

An ice tapper is another antiquated device for cracking ice (although not in bulk quantities). It is a hard metal oval at the end of a plastic handle (pictured in fig 4.28). To use it, hold the ice in one hand while tapping on it with the tapper. Tap lightly and repeatedly rather than with a hard whack: it works better that way, and you won't have a sore hand at the end.

Or you can smash your ice from cubes into smaller sizes using a mallet and a kitchen towel or a canvas bag called a Lewis bag. Wrap your ice cubes in the towel or insert them into the bag. Whack it with a mallet, muddler, rolling pin, meat tenderizer, or something like that. The fabric of the towel/bag will absorb any water created in the process.

Then dump the ice into a plastic bag for storage. If the ice sticks together in the freezer, it's usually easy to break it up again by dropping the bag on the counter or floor.

Like white ice made from coconut water or carbonated water, you can use crushed ice to create a backdrop for bright-colored drinks like the Queens Park Swizzle. This type of ice is useful for drinks that have a "float" of bitters (or another type of liquid like dark rum) on top of the drink and in drinks with flower and herb garnishes.

You can also make colored or flavored ice and then crush it. Coconut water ice would be great in the Queens Park Swizzle, for example. I have made crushed ice from butterfly pea flower cubes too—Mint "Bluelep," anyone?

For cobbler-style ice, rather than crush it from larger ice, I use ice trays that make mini-spheres and squares. I currently have three different size trays: one makes ice spheres about 1 inch (25 mm) in diameter, the other produces smaller, ½-inch spheres, and then I have trays that make ⅜-inch (10 mm) square mini-cubes.

Sherry Cobbler over Cranberry Ice

3 oz (90 ml) fino sherry such as Tio Pepe
.5 oz (15 ml) orange juice
.25 oz (8 ml) simple syrup

Make cobblestone ice with cranberry juice. Mix this ice with unflavored cobblestone ice and frozen blueberries. Mix liquid ingredients. Fill highball or Old-Fashioned glass with flavored ice and pour liquids over it. (Traditionally a Sherry Cobbler is made with amontillado sherry.)

Garnish: mint sprig.

Facing top left, **Figure 4.28.** Ice tapper. This retro-style tool breaks up ice into smaller pieces.

Facing bottom left, **Figure 4.30.** Cobbler-sized ice, some colored with food coloring.

Facing top right, **Figure 4.29.** Trays to make cobbler-sized ice.

Facing bottom right, **Figure 4.31.** Sherry Cobbler with cranberry-flavored ice.

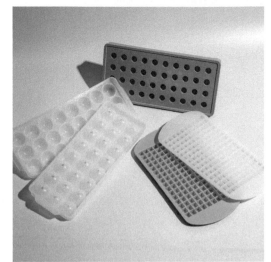

The size and shape of the ice frozen in these trays makes them perfect for colored and flavored ice. You can add just a dash of flavor by throwing a few juice cobblestones in with the rest of the drink, as I did with the Sherry Cobbler in figure 4.31.

Nugget/Sonic/Pebble/Pellet/Chewy/Hospital Ice

This nugget-sized ice is the type that people love to chew on, and is it known by many as "Sonic ice" after the fast-food restaurant. It is about the same size as cobbler/crushed ice, but instead of being broken off from solid ice cubes it is "extruded." Layers of thin flaked ice are packed together and pushed out a tube in small segments.

This ice contains far more air than solid ice does, making it soft and chewy. To make a home version of it, some people suggest freezing carbonated water and crushing it up. I attempted freezing carbonated water in my cobbler ice trays, but it's really not the same experience. You can buy a nugget ice maker for home use, but these are quite expensive.

Shaved/Snow Ice

While snow cone ice is typically broken down or shaved from regular ice cubes, shaved ice (often pronounced without the "d") is shaved off an ice block. Both types of ice are used as the base for cold dessert treats flavored with a range of syrups.

You can make shaved ice by scraping a block with a knife, ice pick, a sturdy fork, or with a specialty ice shaver tool that looks like a little metal box. There are also hand-cranked and electric ice shavers for sale online and in kitchen stores. You can also crush cubes into snow

Figure 4.32. Ice crushing tools: ice crusher, mallet, canvas Lewis bag.

ice: use the same Lewis bag and mallet crushing technique from above, and just keep crushing the ice down into snow.

Shaved and snow ice are used for snow cones, shave ice, and some people like it in Mint Juleps. I think it melts too fast to be useful for drinks, except when it's pressed together and refrozen. There is a classic tiki drink called the Navy Grog that was traditionally served with an ice cone shaped like a buoy with the straw inserted through the middle. That shape is pictured in figure 4.33.

To make it, first create crushed or snow ice. Then press it together in a cone-shaped metal mold, and put it back in the freezer (with a straw inserted for the hole) for a few hours to solidify. You can also use other molds like a sphere mold or an extra-large jelly mold to make other shapes in the same way. Then when you have made your ice shape, you'll need to find a drink to put it in. I bet it will be delicious, whatever you choose.

Figure 4.33. Crushed ice formed into shapes including an ice ball and a buoy.

APPENDIX 1
TROUBLESHOOTING AND ICE STORAGE

My Ice Isn't Perfectly Clear

Whether you're making your ice in a cooler, in specialty trays, or some other way, the reason that ice turns out cloudy when you're trying to make it clear in a directional freezing setup is usually due to the freezer being too cold or the freezer fan blowing on the ice. Let's look at each.

First, it might be helpful to think about *how* your ice isn't clear. If you're using a cooler, is the cloudy part of the ice right in the middle of the block (more likely your freezer is too cold), is there a strange cloudy lump on the surface (fan is blowing on the top), or do you have a lot of little bubbles scattered in the clear part of the ice (vibration issues)?

If you're using a sphere mold on top of a thermos, is the resulting ice ball cloudy in the center (directional freezing isn't working at all, or perhaps the mold sits to high on the insulated thermos), or just near the hole on the bottom (too cold and/or sped up by the fan)? Keep in mind how directional freezing is supposed to work when diagnosing your cloudy ice issues. Think about where your ice is cloudy and what would have to have happened to make it that way.

Figure A1.1. Olive in a clear cube made in a clear ice tray. A metal cocktail pick was used to hold the olive in the middle of the cube while freezing and can easily be removed.

Freezer Is Too Cold

Another way of stating that your freezer is too cold is to say that the temperature is so low that the insulation in your cooler is overpowered, and directional freezing doesn't properly take place. Instead of the water freezing from the top down it freezes from the outside in like an uninsulated ice cube tray. In theory if you wanted to correct for this you could build a better-insulated cooler, thermos, or other ice tray. But the easier way to remedy the situation is to turn up the temperature in your freezer.

Turn your freezer to its warmest setting—which will still be below the freezing point—unless you have a good reason not to do this. The slower the ice freezes, the clearer it will be. The United States Food and Drug Administration (FDA) recommends a maximum temperature of 0°F (–18°C) for food safety, which is far colder than the freezing temperature of water. In my freezer, that's around the maximum temperature it will reach at its warmest setting. So no matter how warm I try to make the freezer, it's still plenty cold.

Another way your freezer can discourage directional freezing is if the cooling element is located directly underneath your cooler or tray. This is the case with the small upright freezer I have at home; the shelves have the cooling elements woven through them. When my cooler or thermos rests directly on the cooling element it overpowers the insulation, and the ice starts freezing from the bottom to the top in addition to top down.

To a lesser degree, if your cooler or insulated tray sits on a wire shelf raised off the bottom of your freezer, more air will hit the bottom than if it were sitting on the solid plastic bottom of the freezer, and you may get less clear ice. This is an easy issue to solve—either move the cooler elsewhere in your freezer, or put something insulated between the cooler and the shelf/cooling element. To improve the situation, I have set my cooler/thermos/ice cube trays atop a plastic cutting board, towel, and even a foam mousepad.

The Freezer Fan is Blowing on the Ice

The freezer fan can make a surprisingly huge impact on the rate of freezing and the quality of the resulting ice. When I received a new freezer a couple of years ago, all my ice started coming out cloudier than the ice made in my previous one. Most of the difference was due to the fans.

The freezer fan can make your water freeze too fast, which happened in my ice ball molds until I set up a wind shield. I draped a resealable plastic bag over the top of the thermos to block the wind. The fan also impacted my large ice blocks: they developed a strange lump on the surface. The wind shield prevented it from continuing to happen—I just rested a plastic cutting board over the top of my cooler.

If you're putting a wind shield on top of your freezing ice in any setup, be sure you're still allowing air circulation. If you seal the surface of a cooler in plastic wrap, for example, the rate of freezing drops to an intolerably slow pace. Allow air circulation but avoid wind.

Too Much Motion

This is a smaller problem than the fan or too-cold freezer. Try to avoid jostling the container holding your freezing water. When you do, the trapped air being pushed down into the bottom of the cooler can form bubbles that float up and stick to the underside of the freezing ice. You may get a small bubble trail, a fireworks-like pattern of bubbles, or even a little pud- dle- shaped pocket in your ice.

The jostling of the cooler can come from vibration if you have a very old freezer that shakes when the compressor turns on. But most of the time vibration comes from members of the household opening and slamming the freezer door. Or the ice gets jostled because you keep pulling the cooler or trays out of the freezer to check to see if they are fully frozen yet. I've done that plenty of times.

For more information about boiling the water or using filtered/distilled water, see "Small Additional Improvements" on page 29.

How to Store Ice

Storing dry ice (ice that is dry on its surface, not frozen carbon dioxide) is easy; you can just put it in a sealed container. Wet ice, on the other hand, can be a bit of a pain.

Wet Ice

If you grab a handful of freshly carved wet ice from the cutting board and put it in the freezer in a bowl, it will stick together surprisingly fast when the surface water of the cubes freezes again. Separating stuck-together ice is sometimes challenging, but at other times it's nearly impossible—the ice sometimes shatters rather than separates. And after going through all the effort to make perfect ice, you don't want it to get glued together. There are a few ways to help prevent your ice from sticking together.

Keep Wet Ice from Sticking Together

- Dry it off. Shake off the water and/or pat the ice dry with a towel or polishing cloth before putting it back into the freezer. This won't prevent it from sticking together entirely but will help a lot.

- Shake and break. This is what I do most often. Put your cubes in a plastic bowl (I use a plastic colander) in the freezer and set a timer for five or ten minutes. When it goes off, shake or smack the bowl so that the cubes come loose from each other. Reset the timer and repeat two to three more times until the ice is dry. Then move it to another container for storage.

- Don't let wet ice cubes touch each other. Or separate your cubes on a cutting board or serving tray when you put the ice back in the freezer. After half an hour or so the cubes should be dry, though they might stick to the tray on the bottom. Pry them off the tray, then store them together in a plastic bag or container.

- Use vodka. Some bar owners who don't have a lot of freezer storage space fill a small spray bottle with vodka and spray it on the ice before stacking pieces together. I'm not sure how well this works.

- Separate the ice into individual bags. I have had to do this when my freezer space was very tight: place each piece of ice in an individual resealable plastic sandwich bag and leave the top open—this allows the ice to dry out without freezing onto other cubes. You can stack these bags on top of each other or throw them to the back of the freezer in between your leftovers. When dry, seal the bags to prevent them from sublimating, or combine them into a container for storage. When you're done, reuse the bags for the next batch.

Dry Ice

Once the ice is made and has dried off so that it won't stick together, keep it fresh in your freezer. Clear ice doesn't become cloudy when you store it in the freezer (people have asked me this), but depending on your freezer conditions the ice may get frosty on the outside. Don't worry, though: the frost will begin melting off quickly as soon as you take it out of the freezer.

Ice shrinks in a typical freezer over time: the ice sublimates, and cubes get thinner. To prevent or reduce this, store your ice in a sealed container. Plastic food containers and resealable plastic bags work great.

The other great reason to seal up your ice is to prevent it from absorbing smells from items in your freezer and refrigerator. This happens surprisingly quickly: one night in the freezer when there is pungent cooked food in the refrigerator is enough to ruin a whole batch.

That's why the crescent-shaped ice from freezers with automatic ice makers often tastes so bad—it is sitting out in the freezer and absorbing both freezer and refrigerator aromas. Transfer the foods in your fridge and freezer to sealed bags or containers, or store your ice in sealed containers—better yet, do both.

Clean Your Smelly Ice Trays

To rid your silicone ice cube trays of freezer/fridge smells, bake them at 350°F (175°C) for an hour. Silicone trays can withstand high heat—but first make sure your trays are silicone and not another plastic that will melt!

Figure A1.2. Three-dimensional ice. Blue-colored ice letters made in a tray stuck to big cubes.

APPENDIX 2
BUYING GUIDE

This list is for specific products that I have used and recommend. If you don't see a product mentioned here that was in this book, it probably means that I bought it online and that there are many versions of it made by several different companies.

Ice Cube Trays

- 2.5-inch (6.3 cm) cubes, spears, 1.25-inch (3.2 cm) ice cube trays from Cocktail Kingdom, CocktailKingdom.com.
- Small cobblestone trays .5 inch (1.3 cm) are from the brand Mydio.

Clear Ice Trays

- Clearly Frozen. Inexpensive tray makes ten 2-inch (5 cm) cubes at a time. ClearlyFrozen.com
- Dexas Iceology. Sturdy trays to make either 2-inch cubes or spheres. Dexas.com.
- Ghost Ice. Industrial quality trays for home and bars that make cubes measuring 2 x 2 x 2.5 inches. GhostIceSystem.com
- SimpleTaste Crystal Clear Ice Ball Maker is a vertically oriented sphere tray and makes 2.4-inch (6 cm) spheres. Multiple retailers.

Ice Ball Press

- Available in two sizes from Cocktail Kingdom.

Ice Tapper

- In a range of colors from Cocktail Kingdom.

Patterned Ice

- The Ice Designer. A tray to make patterns in clear ice. TheIceDesigner.com.

Ice Cone

- Beachbum Berry's Navy Grog Cone Kit from Cocktail Kingdom.

Figure A2.1. Ice-cutting tools.

Nan, Camper, and Allison at the end of the shoot.

CAMPER ENGLISH is a cocktails and spirits writer and speaker who has covered the craft cocktail renaissance for over fifteen years, contributing to more than fifty publications including *Popular Science, Saveur, Details, Whisky Advocate,* and *Drinks International.* After much experimentation, in 2009 he revealed a simple method for making clear ice that is now used all over the world. He has since written dozens of articles and given talks about ice internationally. His previous book is *Doctors and Distillers: The Remarkable Medicinal History of Beer, Wine, Spirits, and Cocktails.*
www.Alcademics.com

Photographer **ALLISON WEBBER** is a former bartender specializing in cocktails and spirits photography.
www.allisonwebber.photography

Stylist **NAN ALLISON** is a multidisciplinary designer and stylist living in Portland, Oregon. www.nan-made.com